The Japanese Experience in Technology:
From Transfer to Self-Reliance

Technology Transfer, Transformation, and Development:
The Japanese Experience
Project Co-ordinator, Takeshi Hayashi

General Trading Companies: A Comparative and Historical Study, ed.
  Shin'ichi Yonekawa
Industrial Pollution in Japan, ed. Jun Ui
Irrigation in Development: The Social Structure of Water Utilization in Japan,
  ed. Akira Tamaki, Isao Hatate, and Naraomi Imamura
Role of Labour-Intensive Sectors in Japanese Industrialization, ed. Johzen
  Takeuchi
Technological Innovation and Female Labour in Japan, ed. Masanori Naka-
  mura
The Japanese Experience in Technology: From Transfer to Self-Reliance,
  Takeshi Hayashi
Vocational Education in the Industrialization of Japan, ed. Toshio Toyoda

The United Nations University (UNU) is an organ of the United Nations established by the General Assembly in 1972 to be an international community of scholars engaged in research, advanced training, and the dissemination of knowledge
related to the pressing global problems of human survival, development, and welfare. Its activities focus mainly on peace and conflict resolution, development in a
changing world, and science and technology in relation to human welfare. The
University operates through a worldwide network of research and post-graduate
training centres, with its planning and co-ordinating headquarters in Tokyo,
Japan.
  The United Nations University Press, the publishing division of the UNU, pub-
lishes scholarly books and periodicals in the social sciences, humanities, and pure
and applied natural sciences related to the University's research.

The Japanese Experience in Technology: From Transfer to Self-Reliance

Takeshi Hayashi

 United Nations University Press

The United Nations University project on Technology Transfer, Transformation, and Development: The Japanese Experience was carried out from 1978 to 1982. Its objective was to contribute to an understanding of the process of technological development in Japan as a case study. The project enquired into the infrastructure of technology, human resources development, and social and economic conditions and analysed the problems of technology transfer, transformation, and development from the time of the Meiji Restoration to the present. The research was undertaken by more than 120 Japanese specialists and covered a wide range of subjects, including iron and steel, transportation, textiles, mining, financial institutions, rural and urban society, small industry, the female labour force, education, and technology policy.

This volume constitutes a broad overview of the interaction between technology and development in Japan since the Meiji period.

United Nations University Press
The United Nations University, Toho Seimei Building, 15-1 Shibuya 2-chome, Shibuya-ku, Tokyo 150, Japan
Tel.: (03) 499-2811   Fax: (03) 499-2828
Telex: J25442   Cable: UNATUNIV TOKYO

Typeset by Asco Trade Typesetting Limited, Hong Kong
Printed by Permanent Typesetting and Printing Co., Ltd., Hong Kong

HSDB-36/UNUP-566
ISBN 92-808-0566-5
United Nations Sales No. E.90.III.A.4
04500 C

Contents

Preface

The United Nations University asked the Institute of Developing Economies in Tokyo to analyse what made it possible for Japan, once an importer of foreign technology, to become an exporter of its own technology.

What prompted the University to propose this was no doubt its realization that the solution of current development problems in the developing countries is a matter of global importance and that, for the development of each, technology transfer leading to self-reliance is necessary.

In addition, to delve into the history of industrial technology in modern Japan from the perspective of the development problem presented us with an interesting academic challenge. One principal difficulty, however, was also anticipated: the current thrust of comparative technological studies in Japan, in both academic and business circles, is overwhelmingly centred on comparison between Japan and the West, ignoring the third world.

Although not all the difficulties faced in our project, with its emphasis on field studies, were overcome, we were fortunate in securing the ready co-operation of more than 120 experts from all parts of Japan in various fields of industrial technology. The project involved a year of preparatory work and another year of editorial work, and the Institute of Developing Economies covered all expenses for several staff members assigned to the project. The contribution from the United Nations University went to research activities and also to building a network both within and outside of Japan.

More than 120 interim papers and 20 volumes of reports were produced during and after completion of the project. Of the reports, 9 volumes have already been published—in Japanese—by the United Nations University and have been well received. Moreover, besides the present volume, a single-volume English translation-adaptation of two of the Japanese reports has been published by the UNU (*Vocational Education in the Industrialization of Japan*). The remaining reports—10 of which have been translated into English—await publication.

vii

This book is a final report of the whole project on the "Japanese experience." The author served as the project co-ordinator, and though a great part of the information has depended on the project reports, this is not a summary of those reports. Instead, it is an independent work, although its subject and that of the reports supplement each other. Nevertheless, the author's views do not necessarily coincide with those expressed in the reports.

Since the aim of the present book is to provide materials for our "dialogue" with those in the developing countries who are responsible for the planning and administration of national development, in chapter 1 I have added an outline of the history of Japanese industrial technology after World War II, which had not been taken up in our project.

To facilitate more effective dialogue, I also present a theoretical framework: the five Ms constituting technology and the five stages of development, from technology transfer to self-reliance in technology. This framework is useful in explaining the roles played by the four chief industrial branches (iron and steel manufacturing, railways, mining, and textiles, or five if we add the shipbuilding sector, which became a supply source of domestically produced machines that could be used in mining and other branches) in the early years of industrialization in Japan and the links between them and industrial policy.

Our project was unable to cover such important industrial branches as the electrical industry or earthquake-resistant construction. Later research by this author found that the role of the food industry was also important. The OEM (original equipment manufacturer) system, commonly employed in today's electrical and machinery industries, had already been established in the food industry more than 100 years ago. This fact could have bearing on our understanding of biotechnology.

Although Japan was fortunate in having such native technologies, some 60 years were necessary for it to accomplish the first stage of industrial revolution, through technology transfer, and to form a national technology network. In the 1920s, the aforesaid four chief industrial branches and the hydraulic power industry were able to establish links among themselves on a minimal scale and at the lowest level, ensuring a "point of no return." The last to enter that process was the chemical industry, which benefited from the delaying of imports during World War I, followed by rapid growth of the machinery industry in the 1930s after the Great Depression.

Besides the external conditions favouring it, the Japanese economy was able to rehabilitate itself so rapidly in the face of the devastation caused by World War II because four of the five Ms were already in place at notably high standards.

Only after the 1970s did the links among the Ms grow deep enough and wide enough for Japan to become a genuine technology-exporting country, but even before that, Japan had won the world's top position in some technology sectors. It took Japan 120 years to attain high technological development, but that was only half what the West had required. And it is likely that Japan's Asian neighbours will require even less time, say half what it took

Japan, to catch up. An industrial revolution through technology transfer is possible, even if beset with sizable difficulties.

It has often been misunderstood that the latest technologies would prove the ones most appropriate for national development. In Japan's case, it did not always adopt only the latest technologies. When it did, it did so because that technology was considered to be the most appropriate.

It should be noted that the choice to transfer the latest technology is possible only after a country has attained primary self-reliance in technology, at which time it should then purchase only the necessary technological systems, ones without high social costs and conflict.

Self-reliance in technology does not mean autarky. Today's technologically advanced countries are not autarkic in technology. Self-reliance in technology refers to the ability to absorb all needed technologies, and the attainment of this self-reliance is accomplished not at a stroke but in stages.

The more advanced a technology is, the wider its links will be and the higher the level at which it will establish itself as a working engineering system. In other words, advanced technology requires high-level, intricate links between itself and technological pre-conditions and between itself and related services. This requirement explains the reason the technology gap between the North and South tends to widen. The question of utmost importance to a developing country is, therefore, how to form its national system of technology at a minimally effective scale and level.

Since its first stage in acquiring technological self-reliance Japan has been dependent on foreign countries for most of its raw materials, and this characterizes its development in technology. Technological development follows a spiralling rather than a straight path, and the question as to which sector of technology a country with a specific resource position may decide to start with is a matter of national consensus.

There is the "textile first" theory of industrialization, but, while it may describe the approach taken by the industrialized countries, it is not necessarily the path of development other countries should follow. A country may well begin its industrialization with power development, food processing, or communications and transportation. The conditions under which industrialization begins cannot be the same for every country; the only common element required is that a national consensus be formed.

One reason Japan could successfully absorb foreign technology in the nineteenth century was because most of the technology and machinery in those days consisted of an assemblage of assorted technologies. Some machines could be dismantled into separate components, and these components could be replaced by parts produced locally through the traditional skills of carpenters, blacksmiths, stone-cutters, metalworkers, etc. Although replacements were often less efficient and poorer in quality, they satisfied national needs in being less expensive and easier to maintain. The repetition of this process made it possible to eventually turn old technologies into new ones.

As with language, imitation is an important step in learning. This step can be very much enhanced by making a thorough examination of traditional skills and technologies. Regrettably, this is not being done in many of the developing countries, where even basic data on meteorological, geological, and hydrological conditions and on natural resources are incompletely available. This is an area where international co-operation would be both useful and necessary, but even then, full use of the empirical knowledge of the local populace should be made.

Toward the end of the nineteenth century, Japan climaxed a decade of trial and error in the spinning industry by catching up with India, then an advanced country in this technology. This was possible because in Japan the basic spinning process was subdivided. After workers had acquired the skills of one process, they were transferred to another. This was an unusual but effective on-the-job method of developing worker skills. Although it proved successful in Japan, it may not elsewhere, especially in countries where the system of technology management is largely based on functionalism and where job-hopping among workers and engineers is common. Further, in Japanese industry generally, this style of training has been combined with the distinctive qualities of the Japanese engineer and with such practices as lifetime employment, rare in other societies.

The point is that every country must find its own way of development. However successful the Japanese experience may appear to be in the eyes of other nations, it was an experience unique to Japan and not one for other nations to follow to the letter. All the Japanese can do is attempt to answer questions about its success and contribute something to the information developing countries need as they search for their own development.

Our project differs in approach from conventional studies, and we have given consideration to such areas as vocational education, general trading companies, cottage industries, and problems of pollution. We hope the information provided here will lead the reader to the more detailed information contained in our individual reports.

Our project on the Japanese experience was first proposed by Professor Mushakoji Kinhide, vice-rector at the United Nations University, and the University's Dr. Uchida Takeo contributed much help as the project moved along. In bringing out this English edition, particular mention must be made of the good offices of Mr. Noguchi Noboru in getting the Grant-in-Aid for Publication of Scientific Results from the Ministry of Education, Science and Culture. The author's deep gratitude is due to these persons and to all those concerned with our project. It is his great pleasure that this work is being brought to the attention of English-language readers.

Also, for much of the work on which the book is based, as with the other books in this series, the author owes a great deal of thanks to his colleague, Professor Tada Hirokazu, and to others at the Institute of Developing Economies who worked under his co-ordination in bringing the project to fruition. The translation into English was done by Mr. Yamauchi Takeo and Mr. Nakai Masao, the author's former colleagues at the Institute. Individual names of others involved are not given here, but the author is in deep gratitude to all of them.

Takeshi Hayashi
Project Co-ordinator

Part I

Overview

Development and Technology in Post-war Japan

Japan in the World

Japan's share in the total GNP of the world was 9.0 per cent in 1980, a position exceeded only by the United States and the Soviet Union.

Because at the beginning of the twentieth century Japan accounted for a mere 1 per cent of the world's total GNP, compared with 30 per cent for the United States and 20 per cent for the United Kingdom, this rapid structural change, and the Soviet Union's rise to second position, are remarkable. The changes in the scope and the structure of the world economy are readily apparent in the 1980 shares of world GNP held by the United States and the United Kingdom, 21.9 per cent and 3.6 per cent, respectively.

In terms of per capita GNP, Japan has achieved a level comparable to that of both the United States and the United Kingdom, inasmuch as its population is slightly more than half that of the United States and slightly less than that of the United Kingdom. In other words, over the past 80 years, the Japanese economy has grown 30 times as fast as the US economy and 20 times as fast as the UK economy. However, this is merely a matter of flows; in stocks, it should be noted, unfavourable gaps remain for Japan compared with either the US or the UK, the latter especially.

With regard to the power of a nation to influence the international community, the United States and the United Kingdom are in a far better position than other nations because English is a nearly universal language. The Japanese language, on the other hand, is not even treated as an official UN language. Thus, when it comes to the question of a country's international political influence, its economic power is not always the decisive factor; this is obvious in the examples of China and India.

Taking population as a criterion, a country with a population of more than 100 million may be regarded as big, but Japan has barely enough population to enable it to count itself among the big countries. Even the United States and the Soviet Union are far smaller in this regard than China and India.

A country with less than US$10,000 per capita national income and less than 100 million population may not be expected to make effective use of a full set of modern technologies because it cannot realize economic efficiency at a level these technologies would require.

Judged, then, in different aspects, Japan may fall outside the group of front runners, but it may be inappropriate to place it among the second-group runners considering the great distance between the two groups. Seen in terms of its industrial power and its governmental system, Japan is Western, but culturally it remains Asian.

Beginning in the 1960s and continuing for more than a decade, the Japanese economy was able to achieve what was then called a miraculous annual growth rate exceeding 10 per cent. Though this was in many ways ascribable to the previous low level of its economic development and to the nation's recovery from World War II, it also reflected the rapid expansion of the scope of production through technology transfer.

Worth noting here is the difference between Japan and the other industrial countries in how it coped with the oil crises of the 1970s, an epochal situation in contemporary history that threw most of the world into hard times. Whereas most countries viewed the crises as a stoppage of the oil supply, Japan saw them as signs of the need to rationalize through technological innovation.

When the economies of the industrially advanced nations were confronted by stagflation, and the United States, which had led the post-war world, suffered a growth rate that had declined to as low as 3.5 per cent (the EC countries had an average of 3.1 per cent), Japan managed to maintain a growth rate not lower than 5 per cent. By the end of the 1970s, much to the perplexity of the Japanese, the world looked to Japan and West Germany to play the role of locomotive, to pull the world economy out of its recession.

It is beyond my ability to fully answer the question of how Japan managed to surmount the crises of the 1970s. One answer that has been offered relates the Japanese success to its capacity for technological innovation, and without doubt, technology has contributed much to the high economic growth rates of Japan since the mid-1960s, a ratio of contribution calculated at 30 per cent. Just as it managed to tide over the oil crises that had brought the high-growth period to an end, Japan also managed to overcome the difficulties caused by industrial pollution that emerged in the 1960s and 1970s by developing technologies to control or prevent pollution and others to conserve energy. These accomplishments brought world recognition to Japan as a technologically advanced country.

Is Japan the front runner of the developing countries, or is it running on the heels of the developed countries? It may be that it has elements of both. In some technologies, though, it is without doubt a leader.[1]

From the time we undertook this project, and especially since 1980, an unusually keen world-wide interest has centred on technology. It seems that the second oil crisis, in 1979, and the ensuing economic difficulties compelled many countries to seek technological innovation as a way to change the status quo.

Something that made it less difficult for Japan than other industrial nations to cope with the oil crises was that industry largely accounted for Japanese oil consumption, thus relegating that portion used by individuals to a less important position than in other countries. This made it easier to develop energy-saving technologies and possibly easier to implement them with more resounding effects. Yet no one can say for certain that technology will be able at all times to play the lead role as a problem solver, as perhaps it has until now.

Indeed, technology alone has not the power to solve economic and related problems. Managerial skills are absolutely vital, as the Japanese experience shows; at the same time, Japan's strategy must be acknowledged as a general solution and not one that is peculiarly Japanese. Thus, it could be said that the Japanese solution is merely one form of the general solution. There have been some studies that pursue this perspective, but we need to examine the question further before coming to any conclusions.

Although technology is not all that counts, its importance is undeniable. In this context, it is not surprising that Japanese technology, with its peculiar history of formation and its unique structure, should have aroused interest among other nations. It is with this in mind that we decided to study the problem.

Our conception of technology and development may differ from the usual. While science is universal, technology is not. What may be called the internal and external links of technology cannot be broken when innovation occurs. In other words, although the internal logic or built-in mechanism of a technology is autonomic, the external conditions under which it must operate are not. Herein lies the dilemma of technology.

Economy and Technology in Post-war Japan

With the world's mining and manufacturing production index for 1975—the year after the oil crisis hit—given as 100, the corresponding figure for Japan in 1980 was 124. By 1980, the economies of all the industrialized countries except Japan stagnated, and the index for the United Kingdom fell below even the 1975 level.

The first to recover from this crisis was Japan, its corresponding index scoring 142 in 1981, followed by the United States (128), France, and West Germany. In terms of per capita GDP in 1980, ignoring the oil-producing countries of the Middle East with figures as high as US$30,000, the Japanese figure, at US$9,890, was 61.8 per cent of the Swiss figure and 89.9 per cent of the US figure. This placed Japan seventeenth among all countries (though fifteenth in 1975). Japan has the smallest personal income gap between rich and poor.

To give a fuller picture, we must consider that Japan depends on imports for 95 per cent of its energy consumption, for 90 per cent of the important raw materials for its manufacturing and mining industries, and for more than 60 per cent of its food requirements. It must be said, therefore, that Japan, though often called an economic superpower, is a vulnerable power—even a minor power in respect to natural resources—a nation that has no other choice but to keep itself going on the basis of technology and foreign trade. Despite the high economic figures for Japan in terms of flows, the livelihood of its people, if not poor, is still far from being rich if seen in terms of stocks. A European Community leader once aptly commented that the average Japanese is "a workaholic who lives in a rabbit hutch."

Even so, the Japanese living standard, not well-to-do but not badly off, is something enviable for people in the third world. The Japanese may live in rabbit hutches, but in the third world even a small dwelling would be satisfactory if clean and sanitary and supplied with tap water and electrical home applicances. For many people in the third world, beset with chronic underemployment or latent unemployment and lacking decent homes, Japan could be a not-so-far-away goal at which to aim. Note too that Japan grew nearly to what it is today in not much more than a quarter-century.

While Japan scored 124 in the mining and manufacturing production index in 1980, the Republic of Korea registered 210. Obviously, the movement of the production index, like that of the growth rate, has no direct bearing on amount in absolute value. The smaller the absolute value of production, the greater the index movement might be, and conversely, the greater the absolute value of production, the smaller the index movement. The continued rapid economic growth of post-war Japan indicates that, because of the great war damage the country suffered, its economic reconstruction had to start from limited, but deliberate, activity and a low level of living.

Post-war Recovery

The cities of Hiroshima and Nagasaki were each destroyed by a single atomic bomb. A great many Japanese cities, with the well-known exceptions of Kyoto and Nara, ancient capitals of Japan, suffered from bombing: in the 119 cities bombed, 2.2 million houses (about 20 per cent) were destroyed and 9 million people made homeless. Because few new houses were built during the war, in post-war urban Japan more than one family—sometimes several —would be jammed together into a house that was already past its prime.

The devastation affected everything connected with daily life, from factories, roads, bridges, electric lines, and waterworks to schools, hospitals, and communications systems. About 40 per cent of civilian national wealth was lost, and the few machines and pieces of equipment that survived were overused, poorly maintained, and short of parts and accessories.

For several years after the defeat, the nation's standard of living hovered at a level of 30 per cent of the top pre-war (1935–1937) level; mining and manufacturing production in 1946 stood at a mere 6.6 per cent of the pre-war high. The greatest losses were in shipping: from a total tonnage of 6.3 million, only 1.53 million (or 24 per cent) had survived.

The railroads were more fortunate, with track loss at 50 per cent and rolling stock loss at a mere 10 per cent, and hydroelectric power plants had suffered only slightly. But with 6 million Japanese being repatriated from overseas and with the presence of the Occupation forces, whose requirements had priority over everything else, the capacities of these two sectors, even if fully worked, could not meet the demand.

Before and during the war, Japan had been largely dependent on Korea for its supply of rice, beans, iron-ore, and anthracite; on Taiwan for rice and sugar; on Sakhalin for timber, wood-pulp, and coal; on Manchuria for iron-ore, coal, and soya beans; and on China for salt, iron-ore, and coal. The stoppage of their supplies as a result of the defeat badly affected Japan's mining and manufacturing industries, and the people suffered from a great shortage of daily necessities.

Extremely short in supply were textiles, with production at merely 33 per cent of the pre-war high; ammonium sulfate was at 42 per cent, paper at 46 per cent, and bicycles at 20 per cent. And manufacturing came to a halt after raw materials were exhausted. The shortage of goods went hand in hand with inflationary spirals.

Intending to materially disarm the militarist-fascist state, the Allied victors prepared a plan toward the end of 1945 for "reparations in kind" to be imposed on the defeated nation. This called for removing or dismantling 50 per cent of the machine tools, all manufacturing equipment of the light-metal and ball-bearing industries, 20 shipyards and naval arsenals, and all plants having a capacity to produce more than 2.5 million tons of steel (the total steel-producing capacity of Japan was 11 million tons). More than 1,000 plants were designated for reparations.

Industrial capacity left untouched at the time was meant solely to produce goods for reparations. It was intended that Japan would revert to a small agricultural nation governed by what Westerners then understood as Asiatic standards; it was to be kept at the level at which it had stood immediately after the 1929 slump.

In other words, Japan should never again rise above the levels of the Asian countries it had trampled underfoot by armed aggression. Its annual production of crude steel, for instance, was not to exceed 1.5 million tons, a level at which it had stood 20 years earlier (1926), and its production would rely solely on domestic ores.

The year of defeat happened to coincide with a very bad rice crop, the second worst in this century, which was further aggravated by typhoons and floods. The rice yield dropped to 60 per cent of an average year, and fears were strong that 10 per cent of the nation's 80 million population might die of starvation. Even through a food rationing system, the Japanese government could not ensure a per-capita daily intake of 1,300 calories.

One observer, an American journalist arriving in Japan at the end of 1945, described the aftermath this way:

The closer we came to Yokohama, the plainer became the gravity of Japan's hurt. Before us, as far as we could see, lay miles of rubble. The people were ragged and distraught. . . . There were no new buildings in sight. The skeletons of railway cars and locomotives remained untouched on the tracks. Gutted buses and automobiles lay abandoned by the roadside. This was all a man-made desert, ugly and desolate and hazy in the dust that rose from the crushed bricks and mortar.[2]

One scholar referred to the Gayn descriptions as a record of a situation characterized by "great heaps of useless war equipment laying about, with throngs of people running pell-mell for the few scraps of consumption goods that remained."[3]

Raw materials could not be imported, and a shortage of fuel greatly hampered transportation. As for electricity, voltage was so low that lamps barely shone. As the currency lost popular confidence, economic life became one based mostly on exchange and barter. A state of marginal existence under rampant inflation from an extreme shortage of goods lasted more than three years. The people were in constant lethargy. A judge, believing that "a bad law is still a law," refused to buy food on the "unlawful" black market and died of malnutrition in October 1947.

Priority Production System and the Dodge Line

Despite the economic difficulties, there were some improvements: The Occupation authorities steadily effected measures to demilitarize and democratize the defeated nation. The emperor myth was unveiled, and the forces that had operated under the aegis of the "inviolability of the Imperial prerogative" were politically ostracized. Women were enfranchised and workers given the right to organize. The education system was reformed. The special political police organization was dissolved, and freedom of speech and freedom of the press were assured.

One of the most important reforms was the land reform, which swept away the semi-feudal landlord-tenant relations. In the three years after 1946, a total of 1.87 million hectares, or 81 per cent, of tenant land, and 240,000 hectares of pasture-land were released from landlord ownership. Most tenant farmers became owner-farmers, with the maximum of landownership set at 1 hectare, excluding some provinces and forest land. Land reform was fundamental in expanding and deepening Japan's domestic market.

The House of Peers, whose membership had been restricted to high taxpayers and absentee landlords, was abolished. This collapsed the material foundation of the ultra-conservative forces that had been opposed to all reforms on the strength of the "inviolability" of the emperor. There are several reasons that explain the quick and successful execution of the land reform.

First, it was done under orders of the Occupation forces; second, the new farmers' unions throughout the country were a force to prevent landlords from sabotaging the reform; and third, landlord rule over tenant farmers had been on the wane through the war as economic controls such as fertilizer rationing and the rice delivery system were imposed. Also, since the 1910s, when tenancy disputes began to be frequent, the government had posted kosaku-kan (officials in charge of tenancy relations) with police power in all prefectures. The kosaku-kan had kept detailed accounts of the tenancy disputes they had handled, and these records were helpful in reform adminis-tration.

During the time of the reforms, the economic life of the nation, aggravated by inflation, showed no signs of improvement. A plan was drawn up to give priority in recovery to the basic industries, namely, steel, coal, fertilizer, gas, cement, and railroads. Under this plan, the "priority production system," labour and money were first to be put into coal-mines; then coal was to be produced for manufacturing iron and steel, and the steel materials were to be used for increasing coal production. It was hoped that in this way allied in-dustries and others would be stimulated and the inflation resulting from the shortage of goods would gradually be overcome.

This recovery plan, though theoretically reasonable, was misguided. To begin with, the existing coal-mines had obsolete, worn equipment whose maintenance had been neglected in the wartime drive for more coal. Skilled miners were in short supply, 20 per cent of the total being inexperienced. Three to five years would be necessary before many of the mines could re-cover their pre-war levels of output. The annual coal output per miner was only 90 tons, versus the pre-war (1930–1934) average of 200 tons.

Second, although daily-necessity consumer goods were in extremely short supply, the demand for steel and other basic producer's goods was not great enough for their manufacturing capacities to operate profitably or for the labour force to be effectively employable. Hence, their market prices had to be even lower than their production costs. The government, therefore, sub-sidized these industries to cover the backspread.

Since the steel industry was more capital-intensive than coal-mining, it could recover faster than mining when supplied with imported raw materials and subsidized by the government. The priority production policy thus stimu-lated recovery in these industries, but it did not eliminate inflation.

Priority was also given to the increased production of ammonium sulfate fertilizer, needed for rice cultivation. As symbolized by the 1946 "Food May Day" demonstrations, the food shortage was an important part of the critical economic conditions and a key factor in the political and social unrest at the time. The government therefore treated the chemical fertilizer sector with special political care, and by 1949 it had recovered its pre-war level of production.

Although the estimated requirement of steel materials for use in coal-mines in 1946 was 98,000 tons, only 80,000 were allotted, of which 25,000 were illegally disposed. Only slightly more than half the required steel, there-fore, was put to use in the mines. The Occupation authorities ordered the Japanese government to make available 2 million tons of coal monthly for the people, but the government was hard pressed to raise its target level even to 1.2 million tons. The actual monthly output of coal in November 1945 was only 554,000 tons.

The situation regarding cement was no better. Under the cement distribu-tion sytem, at least 70 per cent of the requirement was to be made available, but what actually appeared was less than 50 per cent. Workers often blamed management for sabotaging production by concealing and illegally disposing of goods and materials. Struggles of the newly legalized labour unions sometimes even led to worker control of production.

A strong distrust of the management running the mines, in which a vast amount of state funds were invested, clouded their operations. This distrust was clearly evident in the proposal by the British representative on the Allied Council for Japan that the state take control of *zaibatsu* coal-mines for three years. There were even apprehensions about entrusting to the private sector the nation's post-war rehabilitation. The proposal for state control of the coal-mines was finally abandoned after a frantic resistance by management. And management soon regained control of the mines where production had come under worker control.

By 1949, thanks to the government's emergency aid in addition to the intended effects of the priority production system, industrial production had largely caught up with inflation. Then, however, the Occupation authorities ordered the Japanese government to change its policy: first, economic aid to Japan was discontinued; second, the price-offsetting subsidies were ordered discontinued; third, a balanced finance policy would be taken to cope with inflation; and fourth, Japan was brought back into the international economy by the introduction of a single exchange rate of US$1 : ¥360.

With this policy change, the coal industry, which had been allowed to operate with an overemployment of labour to increase coal output, was now compelled to raise its productivity and therefore to rationalize and mechanize its production system. It was imperative now not merely to produce more but also to realize lower prices through increased productivity. The steel industry and other basic industries, which, with the aid of government subsidies, had been able to buy coal for ¥1,000 a ton, were now forced to pay ¥3,344 a ton. These high prices formed a bottle-neck that impeded economic reconstruction.

As a part of the mechanization in the coal industry, coal diggers and loaders were imported from the United States with aid funds. But with pit conditions, coal-beds, and other production conditions being much different from those in US coal-mines, they were soon found awkward to handle and left unused. It is easy to see that this early case of technology transfer failed because of the casual handling under foreign aid. But it is important to note that the unusableness of the American machines (even though this pushed up the price of coal) gave impetus to manufacturing the machines domestically. Because the machinery industry had been very much munitions oriented during the war, it found itself in need of new markets in this period; consequently, the coal industry was a welcome customer.

The Japanese coal industry next turned to Europe for the necessary technologies, and in 1950 it introduced the Kappe method of coal-mining from West Germany. By the following year, this technology had begun to be adopted by the leading coal-mines, and, coupled with the successful development of shafts that had been in progress in some of the major coal-mines, it raised productivity. Coal output approached 50 million tons in 1951, and productivity became comparable with the levels of most European countries.

The steel industry began peacetime work with three operating blast-furnaces at the Yawata Ironworks (in its heyday, the industry had had a total of 37 blast-furnaces). The newest of the nation's furnances (affecting 22 plants, or the equivalent of three-fourths of total capacity) were designated for reparations, most chief executives were purged, and the biggest of the enterprises was dissolved under the economic democratization policy of the Occupation authorities.

The recovery of steel production was slow, but after a mere 560,000 tons in the year of defeat, it recovered four years later to 70 per cent of the pre-war level (or to 4.84 million tons in crude steel). Then came the government's abrupt changes in economic policy and, like the coal industry, steel suffered a serious blow. Although it had succeeded in introducing a technology enabling it to use ordinary coal instead of raw coal, the steel industry could not achieve marketability without the aid of price subsidies.

As we have seen, the government's abrupt policy change dealt a serious blow to the recovering economy. Major corporations were forced to dismiss their employees on a massive scale, and some were even driven to bankruptcy. The government's reduced budget policy came suddenly, at a time when industry had not yet managed to fully recover productivity and when many enterprises were unable to meet market needs because production costs were too high. The new policy, the so-called Dodge Line policy, quickly ended inflation, but it increased uncertainty about the future of the Japanese economy.

The strategy for economic recovery based on coal and steel thus had to be discontinued, and priority was shifted to shipping, electric power, and transportation. Of all branches of the economy, shipping had suffered most, and if Japan were to be brought back into the world economy, the recovery of this sector was urgent.

But a more important reason for a priority shift to shipping was that the Occupation policy, which had designated shipping for reparations of a punitive character, was now beginning to change. As the cold war progressed, the United States, which had played an almost exclusive role in the Occupation, was now increasingly in favour of using Japan and West Germany as factories to help rehabilitate their respective neighbouring countries. Also, many US politicians were beginning to feel that if the financial burden on the American taxpayer were to be lessened, the Japanese economy should be made to stand on its own feet.

The Korean War and Japanese Recovery

An unexpected turn of events came with the outbreak of the Korean War in June 1950; it galvanized the Japanese economy back to life. Social reforms that had been dragging amid the chaotic economic conditions began to show progress as the economic life of the nation grew active.

Within the first year of the war, the "special procurements" reached US$340 million; this more than cleared all the backlogs in the manufacturing industries that had been caused by the Dodge Line policy. Goods and materials for use by the UN forces ranged from locomotives, rails, trucks, steel materials, iron posts, electric wire, barbed wire, and other heavy-industry products to chemicals, processed foodstuffs, clothing, and medicines. The procurements reached into all branches of Japanese industry; three branches alone—metals, machinery, and textiles—accounted for 70 per cent of the special procurements.

Covering also the goods and materials for the post-war rehabilitation of South Korea, the special procurements amounted to a total of US$2.4 billion in the four years after 1950, which, even after deducting the cost of imported raw materials, left Japan with a big dollar surplus. The Japanese economy had thus struggled free of its worst difficulties.

The special procurements demanded that Japanese industry mobilize all its existing equipment, however worn and used, so that most of it soon needed replacement or renovation. And this was made possible by foreign currency earnings. Indeed, the first real impetus for Japan's post-war recovery came from the special procurements connected with the Korean War; in other words, the stimulus came from outside Japan.

For example, the steel industry, whose reconstruction based on the priority production system had been stopped by the Dodge Line policy, took advantage of the Korean War to expand its capacity by importing new equipment and realized not only lower prices for its products but also improved quality. What made this possible was the favourable conditions in the international technology market. Technology transfers were very liberal, and Japan's steel industry acted wisely in its choosing and importing of new technology. We will return to this point later in the discussion.

The strip mill is an example of the sort of technology transferred at this time. Compared with older types, it was automated and of far greater speed. Though new to Japan, the technology was already well established in countries with advanced steel industries. Japan had failed to introduce this technology earlier mainly because of the heavy military orientation of the steel industry and because the industry was under state control. The post-war transfer of technology was aimed just as much at the recovery of the steel industry as it was at overtaking the advanced nations.

Another new technology was the basic oxygen steel-making process, also known as the LD process, which was, at the time, the day's newest technology. As an Italian case later reveals, it had not yet been globally established. Nevertheless, the Japanese steel industry adopted and eventually improved the process by adding new ideas and devices, thus laying the foundation for the industry's future development.

Because a strip mill rolls steel in a continuous process at a high speed, mass production became possible. Moreover, the production of high-quality steel sheets had not been possible with the old rolling mills. Thus, Japan was now able to produce materials for use in cars, small electric appliances, and other durable consumer goods, and steel makers could now also mass-produce materials for the general machinery industries. This was all of great significance to the steel industry, which had functioned entirely under the limitations of steel-plates, bar-steel, and section-steel production. Also, with the introduction of LD converters, indispensable for the mass production of rolled-steel products, the two processes of input and output became well balanced. (In most developing countries, they tend to be poorly balanced.)

In another area of the steel sector, a plan for an innovative mill materialized at this time, and the result elevated Japan to a position of world influence among steel makers. Kawasaki Steel Corporation drafted plans for a seaside mill in which the continuous operation of pig-iron production and steel-making was possible. It was a completely new plan both in mill placement and layout. Raw materials (ore and coal) would be unloaded on a wharf at the mill site, undergo manufacturing processes, and emerge as manufactured goods for shipment from another wharf at the same site. At the Yawata Iron Mill—the oldest of Japanese iron and steel works, where a half-century of expansion had meant one new shop or facility after another—the seemingly endless adding-on of the intramill transport railroads extended some 400 kilometres. Plant redesign shortened this by 90 per cent.

Though a change in mill layout may appear to be an insignificant adjustment, when done correctly it can save immense transportation time and fuel costs, which grow in scale as production increases. The result of this amazing foresight soon became status quo as all other steel mills hastened to follow suit.

The idea had been developed during World War II, but the Japanese military had opposed it, and even during the post-war reconstruction, it had failed to materialize. Then came the Korean War, which helped move it from the drawing-board to reality. With the mill's new location and layout, Kawasaki Steel was able to produce 700 tons of steel a day. But there was still some opposition, this time from voices in government circles who felt Kawasaki's transition from a major manufacturer using electric furnaces to one using blast-furnaces might bring on an overproduction of steel. In 1950, Japan's annual output of crude steel had been only 5 million tons.

Overproduction did occur in the 1970s, when the productive capacity for crude steel in Japan reached 110 million tons a year. And a decade later, amid a drop in the world demand for crude steel, Japan's top steel manufacturer, with an annual crude steel production capacity of 50 million tons, had to curtail operations to 60 per cent of capacity.

One of our collaborators in this project, Professor Hoshino Yoshiro, has pointed to several factors that sparked the remarkable growth of the Japanese steel manufacturers, growth that saw the capacity of one soar to 10 times what the immediate post-war output level of all Japan had been.

According to Hoshino, at the time, steel manufacturers throughout the world were competing to enlarge the scope of production, and each country was developing components of technology with little regard for what other countries were doing. Under these circumstances, if a steel manufacturer were observant and could collect data on these various component technologies and integrate them into a single system, he could build the most advanced steel mill in the world. And indeed, Japan at the time was fortunately in a position to fully utilize the advantages of the late comer and ready to spend the time and expense necessary to do this.

This was true not only with steel-making technology but with nearly all other technologies, and here the Japanese experience can serve as an important and useful example. Collecting, examining, and appraising relevant information and bringing it together into a consistent whole should constitute a part of the technological development capability of all the technologically less-developed.

In the third world today, however, several factors make this difficult, if not impossible. These include factors inherent in current technologies and factors relating to the lack or immaturity of external conditions of certain technologies that might enable the less-developed to make use of advanced technologies.

Nevertheless, each developing country must work to overcome these obstacles by setting goals and executing plans based on its particular philosophy of development. Ultimately, development is a matter of national sovereignty.

Post-war Japan had an urgent need to rehabilitate itself, and there was an overwhelming national consensus regarding the indispensability of promoting science and technology through introduction from abroad. There was also the general feeling that Japan's defeat in World War II was due in large part to the antipathy of the Japanese military toward science.

There was a wide range of views, arguments, and counter-arguments in regard to the policies for rehabilitation, especially concerning whether Japan should follow an autarkic line of development or one that would make it an integral part of the world economic system. Throughout, however, a national confidence in science and in democracy prevailed and, indeed, characterized the nation's state of mind in the post-war years before the period of rapid economic growth.

From Recovery to Rapid Growth

Rehabilitation and Technology Transfer

As stated earlier, the Korean War was an unexpected shot in the arm for the Japanese economy, which, before it had managed to rehabilitate itself, was drowning in a stabilization crisis. It gave Japan a springboard for rapid recovery in the 1950s and for rapid economic growth in the 1960s. It may even be said that the Korean War changed the entire outlook of the Japanese economy.

Post-war Japan may be divided into five periods:
1. Post-war chaos (1945–1949)
2. Decade of recovery (1950–1959)
3. Decade of rapid growth (1960–1969)

4. Decade of adjustment (1970–1979)

5. Contemporary uncertainty (1980s)

There are those who contend that Japan's rapid economic growth began with the Korean War, because in the late 1950s its economy had already posted high growth rates, high even on an international scale. An official Japanese document concluded in 1956, only 10 years after the end of World War II, that "the post-war period is over."

Some indicators may in fact justify the belief that the special procurements during the Korean War enabled the Japanese economy to recover its pre-war levels. Under this line of argument, Japan entered the period of rapid economic growth in the latter part of the 1950s, a period that continued until the oil crisis of 1973. A similar view also characterizes the years from the late 1960s to 1973 as a period of uncertainty for Japan, pressured as it was to internationalize its economy.

For my part, however, I do not consider the post-war period to have ended in 1956, as the Japanese government declared. At that time, Japan's per capita national income was only US$220 (less than 7 per cent what it was in the United States and 50 per cent in West Germany); more than 45 per cent of Japan's population belonged to the primary industry sector; and as the special procurements came to an end, only light-industry goods such as textiles and sundries were competitive as exports.

To be sure, some economic indices for 1955 might compare favourably with those for 1930, but in the early 1960s the nation had really only recovered what it had lost in World War II. The 10-year income-doubling programme was officially declared in 1960, by which time full employment had been realized and there had developed a shortage of labour as the economy increasingly internationalized. Also at this time there were official plans for the liberalization of trade and capital transactions.

Technology transfer began to increase rapidly as Japan prepared for the imminent arrival of foreign capital and technology, considered a possible forerunner of another national crisis.

Furthermore, the technology transfers of the 1960s differed from those of the 1950s. Whereas the earlier effort was aimed at recovering pre-war production levels, the transfers of the 1960s aimed to prevent an influx of foreign goods and to strengthen Japan's position in the impending international commercial war in which Japan would be forced to compete. Thus, the enlarged scale of production was for much more than domestic demand, and, moreover, the technologies would be the world's most advanced.

The situation much resembled the one 90 years earlier, when the new Meiji government committed itself to building an industrialized country under the slogans "promotion of industry" and "prevention of imports." The great difference between the two times, however, was that the national consensus in the Meiji period was based on creating a "rich nation and a strong army," while in the 1960s it was restricted to non-military wealth and power.

Thus, technology transfer in the 1960s was characterized not so much by an intention to expand the scale of production, to mechanize and rationalize, as was the case in the preceding decade, but by the aim to transform the production system itself into automated high-speed mass production.

There had been a mass-production policy in the 1950s, at least in some industry sectors, but it did not stress high-speed production, much less automation, because a plentiful, good-quality, labour force was then available, making automation less attractive.

Technology transfer in the 1950s, the 1960s, and the 1970s may be characterized as follows:

1. In the 1950s technologies were transferred to bridge the wartime gap in such sectors as steel, shipbuilding, chemical fertilizers, and textiles, sectors that were already active in Japan.
2. In the 1960s technology was transferred in such fields as automobiles, small electric appliances, and petrochemicals, industries that were already well developed in the United States and in the industrialized European countries, but that were still in their infancy in Japan. As a result of the transfer, these products began to be mass-produced as domestic products and became highly competitive with foreign goods in Japan's home market.
3. Technologies transferred in the 1970s included electronics, high-polymer chemicals, and atomic energy, which had been developed during and after World War II. In these fields—except for atomic energy—and in particular electronics, Japan followed a painful path of quickly overtaking the advanced countries, then being outrun, overtaking them once again and, in some fields, taking the lead.

Even before Japan's international competitiveness in the most advanced technologies had become globally recognized, its steel-manufacturing technology was drawing foreign attention. In 1964, one of the biggest new steel mills in Europe, an Italian steel maker in Taranto, had newly completed construction of two blast-furnaces with a capacity of 2,000 tons each and a converter with a capacity of 3,000 tons. When it encountered problems in its blast-furnace operations, it turned to Yawata Steel for technological advice. Within six months, Taranto had been able to increase its output 15 per cent.

Later, Yawata exported converter technology to British Steel Co., in Wales, the birthplace of modern iron-manufacturing technology. Japan's export of such technology culminated in a series of plant exports to developing countries, including Brazil (Usinas Siderúrgicas de Minas Gerais, or Usiminas), Malaysia (Malayawata Steel Co., Ltd.), and the Middle East (Qater Steel Co.).[4]

In the 1960s, the world steel industry entered an age of large blast-furnaces and LD converters, although these plants were still at the planning level and were not yet practical as operational technologies. Thus Britain, a long-time iron-manufacturing country, had to seek help from Japan. This reveals that the components of a technology are usable only when they comprise equal elements of the technology. The question of whether a technology can be used is determined by the least developed of its components. This is where technology differs from science, which endeavours to uncover a new principle and theoretically build upon it. It is important to keep this difference in mind when discussing science and technology.

Although scientific creativity is directed toward the discovery of principle and theory, technological creativity lies in finding a new way to co-ordinate and direct a set of skills and devices toward a definite practical purpose of operation. In R. & D., or research and development, the R may be expressed as a total of ds: $R = d_1 + d_2 + d_3 + \ldots d_n$.

Japanese science and technology have sometimes been characterized with a small r and a large D, r. & D., but I believe they have both contributed their share to the world's Rs and Ds. All national experiences are different, none being superior or inferior to any other. The evaluation standards for pure science must not be applied to technology, which is for meeting the daily needs of the populace.

The post-war Japanese experience can be summarized by taking micro-electronics as an example. Until recently, the vacuum tube was used in communications and computational equipment. In the mean time, the transistor was invented, just at a time when Japan was taking great pains to improve the performance of vacuum tubes and to mass-produce them. Nevertheless, Sony Corporation introduced the transistor into Japan—the first to do so—from the United States, where the technology had been used mainly for military purposes. After 1956 Sony began to develop and manufacture transistor radios, although they were too expensive then for most Japanese.

By 1960, Japanese transistor radios were finding their way into the American market. The transistor itself, as small as a grain of rice, was easier to put together and required the use of fewer hands to produce than the vacuum tube, but it required intensive labour to attach the reed wires to each tiny transistor, to set the resistor, condenser, coil, and variable condenser, and to run through the complicated process of wiring before a radio was completed.

Thus, "the greater the transistor radio industry grew in scale, the more hands were needed. It was the ideal growth industry for Japan at that time where a comparatively cheap labour force with fairly high technological ability was amply available."[5] Japan thus became the top transistor manufacturing country in the world by around the mid-1960s.

In 1960, Japan took another punch, the IC shock. The US corporation Texas Instruments invented ICs and sold them to the US Air Force, though at the high price of US\$700 per circuit. The switch from transistors to ICs represented also a change in the substrate material, from germanium to silicon. In effecting this substrate change, the Esaki diode, a diode discovered by a Japanese scientist, was used, which in itself indicates the character of Japanese industry. As with the invention of KS magnetic steel by Honda Kotaro in 1933, however, it was not Japanese industry that put it to practical use. Science and technology will not be put to practical use where there is no need; even when a need exists, it might not always lead to a practical application. In any event, Japan had no military need for the new technology at that time.

ICs began to be manufactured in Japan in 1966. As is well known, the IC comes in two types: the bipolar type, which is good at quick calculation but not at minuteness, and the MIOS type, which shows just the reverse characteristics of the bipolar model. Though the bipolar type is preferred for aerospace and military purposes because of its high-speed logic circuitry, the MIOS type was chosen in Japan to develop IC manufacturing for civilian purposes because of its better storage capabilities.

One difficulty Japanese manufacturers were facing at the time arose from the fact that some leading US manufacturers were having their units built in South-East Asia, where cheap labour was available. The Japanese makers knew, however, that if they could double their output through mass production, they would be able to realize a 30 per cent lowering of cost; consequently, they introduced more than six times the number of existing automated lines to rival the US manufacturers.

Then in 1971, Japan was hit by the advent of the LSI. The Japanese-made IC would lose the war. Japanese manufacturers managed to cope with the difficulty, however, by increasing the integration density of IC components by one digit, which resulted in a one-digit cost decrease.

An important factor at the time was a curtailment of manufacturing processes. The existing equipment for SSIs and MSIs, including even what had been installed within the past two years, were scrapped to prepare for LSI manufacturing.[6] This heralded a fierce competition between the ability to develop technology and to manufacture it, a waste, it may be said, of human energy. This sort of battle is being fought even today in areas of product development between Japan and the United States and among Japanese manufacturers.

With the LSI, the efficiency and control of machines was greatly enhanced. Industrial technology, whose products in the 1950s and the 1960s were characterized as "big, long, heavy, and thick," was producing in the 1970s goods that were "small, short, light, and thin."

Due to the complexity of the LSI, manual labour had a limited part in its manufacture. Rather, highly complex equipment was necessary, which required heavy capital investment, and this, in turn, demanded a big market. Japanese LSI manufacturers chose non-military areas in which to sell their goods.

IC manufacturers in the United States tended to be venture businesses with a specialty line, but in Japan the chief manufacturers of instruments had their own IC branches, thus their comparative advantages in capital investment, marketing, and product development.

The Effects of Technology

What will the rapid development of semiconductor technology bring to mankind? One argument made in response to this question at a UN University meeting underlined that the development of micro-electronics (ME) would radically change the information and communications networks in the developing countries. The effects of using ME for educational purposes were also discussed and opposing arguments heard. A situation might arise, it was argued, in which a country's central government would make use of ME for monopolizing information so that central needs might be met at the cost of provincial needs.

Does the LSI signal a new industrial revolution? The arguments began when the IC was first used to operate machine tools (the advent of the numerically controlled, or NC, machine tools). More interest was aroused when the machining centre (MC) made its appearance, followed by robots for welding and painting. The NC machine tool in its early years differed from today's in output and price as greatly as the IC and the LSI did.

In 1980, the NC machine tool almost doubled its built-in capability, compared with its predecessor of a year earlier. The machinery industry prompted the appearance of these automated machines in its call for higher-speed mass production and greater product precision. In the automobile industry, for example, the structure of which is shown in figure 1, each car required about 30,000 parts, constituting 5,000 different types. Even the largest car makers manufactured an average of only 30 per cent of these parts. The rest were supplied by small independent manufacturers, and that is where the need for NC machine tools was felt the most.

The adoption of robots in Japan (initially in the small- and medium-scale industries) to weld and paint was encouraged by a labour shortage and the lower prices of LSIs. The introduction of the robot was, it should be noted, a labour-saving device only at this stage, since the intention was to meet the existing tact of the production line; consequently, there was no time savings or loss.

The appearance of the NC machine tool represented an important innovation for the machine tool in the machinery industry. As noted earlier, machines were changed in the 1960s, and in the 1970s, factory layouts were altered. Furthermore, to be able to handle the new machines well, the workers were required to have the basic mechanical and mathematical knowledge of a technical high school graduate. Today's machines, however, require less ability to operate. It is clear, though, that higher educational standards are a prerequisite for higher technology.

Though the NC and MC revolutionized the parts-manufacturing processes, some 75 per cent of the labour and working hours in the machinery industry were in assembly. Consequently, assembly process automation was the next object of rationalization. The Japanese machinery industry is currently testing assembly automation, referred to as factory automation (FA), and the flexible manufacturing system complex (FMSC). It is estimated that, if such automation can be successfully implemented, labour productivity could be enhanced 30 to 40 times. At the same time, workers would be expected to have the skills of a multi-skilled mechanic, skills far exceeding in quantity and quality what they have today. That would require additional investment for education, whether public or private.

Complex manufacturing systems result in lower prices and diversification. The LSI has already changed the character of the mass-production system,

Figure 1. Division of labour in the Japanese automobile industry

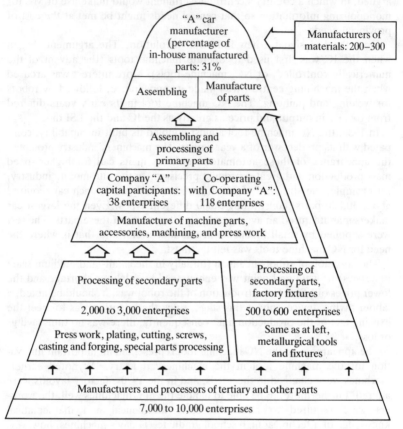

Notes:

1. Percentage of in-house manufactured parts =

$$\left(1 - \frac{\text{Purchasing cost} + \text{Amount paid to sub-contractors}}{\text{Total manufacturing cost}}\right) \times 100\%$$

2. Sub-contracting manufacturers of primary parts do not necessarily work for Company "A" alone.

3. In Japan, parts manufacturers are, as a rule, affiliated with one or more controlling companies; in the United States, they are vertically integrated with automobile manufacturers; in Europe, there is a horizontal division of labour between the two. With Japanese auto makers, despite the comparatively low percentage of parts made in-house (25% to 30% for Japan versus 50% to 60% for the United States and Europe), quality control and cost control are well maintained because they control their sub-contractors' technologies, capital, and personnel.

Sources: Chūshō kigyō hakusho (White paper on small- and medium-scale enterprises), 1980 edition; Industrial Bank of Japan, Research Department, ed., *Nihon sangyō dokuhon* (A reader on Japanese industry), Tokyo, Toyo Keizai Shimpo Sha, 1984, p.163.

allowing, as it does, the production of a uniform product through the assembly of a great many standard and exchangeable parts.

Computers enable mass-production lines to meet the specifications for more than 200 parts that go into the manufacture of a particular car model. Thus, mass production has undergone immense qualitative change, from the mass production of a single kind of Colt rifle (the first mass-produced product) to that of highly diversified products. What has made this possible was the development of the electronics industry in the 1960s and thereafter and the introduction of its products into the manufacturing process.

The development of the electronics industry has caused great concern about how the advances would affect employment. In Japan, this worry has so far proved unfounded, according to an official survey.[7] The appearance of the quartz watch is an example: Technological innovation in one of the manufacturing processes increased productivity four times. But, rather than simply decreasing the number of workers by four times, it was the policy of Japanese business to transfer those displaced to another process. Here we see a great difference between management practices in Japan and those in the United States and Europe, where management is characterized by functionalism.

The enhanced productivity called for an expanded market, and the rapid economic growth of Japan at that time provided it. Without enlarged markets resulting from product diversification, enhanced productivity as a result of technological innovations will reduce employment.

In Japan, increased productivity, a realization of full employment, and wage increases led to an enlarged and deepened market, which proved the government's growth-oriented income-doubling policy effective. As a result, the world-wide reputation of Japanese goods being cheap but poor changed in the 1960s to cheap and good, and since the 1970s they have been regarded as expensive but superb. After IC manufacturing became automated, the cost performance of the Japanese electronics industry began to be highly regarded in the American market.

Besides changing the nature of its products, Japanese industry has now begun to change its employment structure. The total number of employees in the manufacturing sector is on the decline, while in the non-manufacturing sector, especially in sales and in R. & D., it is increasing. In manufacturing, the technology is mature, and the use of ICs and LSIs has generally had a great skill-saving effect, resulting in differences in the quality and efficiency of goods between major and minor manufacturers being almost indiscernable. To use our terms, M_3, M_4, and M_5 have become weightier.

In the case of a certain calculator manufacturer, 1,000 of 3,500 employees are engaged in R. & D. at a technology centre. At a motor cycle manufacturer, salesmen participate in meetings for technological development so that the company's manufacturing technology may better meet market needs, and in the manufacturing department, workers are encouraged to acquire skills not directly related to those required for their current employment. For example, an assembly-line worker may be encouraged to qualify as a maintenance technician, as a plumber, or as an operator of high-pressure machines or instruments. Though this practice may pose a risk to the employer that skilled workers will resign their jobs, it is considered desirable that a single-craft worker should become an all-around worker: hence, the big investment in employee education. Current technological development and innovation require the convergence of a wide range of engineering and scientific knowledge, and, likewise, it is necessary for workers, at all levels, to have a proficiency in several areas, and this represents a new means of skill formation.

It should be noted, however, that the development of technology in Japan was not without cost. First, it widened regional gaps in development; second, it aggravated industrial pollution. The underground water pollution caused by LSI factories has recently attracted attention.

Growth-oriented economic calculations, with their peculiar values, disregard such social problems. If a pollution victim loses all income because of illness, the case will be counted a negative in the economic calculation, but if the victim receives medical care, it will be counted a plus. In the same sense, the building of antipollution facilities will mean an increase in GNP. This should be taken into consideration when one makes use of macro-economic calculations.

Between the periods of recovery and rapid economic growth, Japan's industrial picture underwent radical change. The four major industrial centres of pre-war Japan combined to form a single long belt. This concentration widened the income gap between the urban and rural sectors, which further intensified the concentration of population in the cities, aggravated the urban housing problem, and pushed up land prices to result in the mushrooming of "rabbit-hutch" dwellings. At the same time, the exodus of young people from the remoter towns and villages created areas of underpopulation.

When a community's population decreases below a given level, the community cannot continue to exist; once its working-age inhabitants are gone, its social balance is lost. The phenomenon of village disintegration (*mura-tsubure*) appeared in many parts of the country, brought on by the decline in the primary labour population because of industrialization. This went hand in hand with mechanization, which also contributed to a decreased labour force.

The middle-aged and the elderly, unable to adapt to the changes brought on by the rapid economic growth, were placed at a disadvantage. This may be said to parallel the problems arising in third world countries in their development. Their problems today and the problems facing Japan during post-war industrialization are essentially the same in character and structure.

The pollution problems Japan faced also parallel the situation in the developing countries. In the period of rapid economic growth, which was strongly oriented toward the heavy and chemical industries, little attention was paid to the problem of pollution.[8] Aside from automobile exhaust fumes, the noise and vibration from the Shinkansen (superexpress bullet trains), and the smoke and dust from the growing steel mills, air pollution caused by petrochemical plants gave rise to asthma and related disorders among the

Figure 2. Transactions in automobile production (for every 1 million yen, based on 1975 prices)

	Cars	General machines	Electric machines	Pig iron and crude steel	Non-ferrous (primary)	Rubber	Synthetic resin	Coal products	Petroleum products	Crude oil and natural gas	Trans-portation	Wholesale and retail	Finance and insurance	Services
Automobiles	35.8													
General machines	9.2	3.6								0.5				
Electric machines	4.5	0.7	1.7											
Metal goods	2.0	0.2												
Pig iron and crude steel	8.9	1.4	0.2	13.7										
Non-ferrous (primary)	2.8	0.2	0.5	0.2	1.8									
Textiles						0.4								
Paints	1.3													
Rubber	6.2					0.4								
Synthetic resins							0.8	1.5						
Inorganic basic chemicals							0.4							
Coal products				1.9				1.2						
Petroleum products				0.4				0.5	3.1	0.9	0.3			
Electricity	0.8			0.7	0.4									
Crude oil and natural gas				1.0	1.2									
Transportation	1.6	0.2		0.3							0.2	0.5		
Wholesale and retail	6.0	0.2	0.3	0.5		0.3					0.3		0.3	
Finance and insurance	3.7	0.3	0.2	0.4	0.2	0.3					0.3	0.2	0.5	
Services	2.8	0.6	0.5	0.6		0.3					0.2		0.5	

Note: Numbers in circles correspond to the size of transactions; the structural relations shown here indicate the structure of transactions between different branches directly and indirectly necessary for producing a unit of automobiles.

Source: Prepared by Watanabe Toshio and Kajiwara Hirokazu on the basis of Ozaki Iwao, "Reaction of Economics to Changing Structures: Technological vs. Economic Systems," *Kikan Gendai Keizai*, no. 40 (1980). Taken from *Ajia suihei bungyō no jidai* (Horizontal division of labour in Asia) (Tokyo, JETRO, 1983).

populace in the areas around the plants, and the heavy-metal effluence from fertilizer plants, ingested by fish, eventually culminated in the tragic outbreak of Minamata disease.

It should be noted here that the "experts" in the mercury poisoning cases denied, on the basis of data from oversimplified laboratory experiments, any causal relationship between the probable sources of the pollution and the illness of the victims. Even when they were unable to deny the facts any longer, these "experts" aligned themselves with those responsible for the pollution in minimizing the harm. Political parties and labour unions failed to act effectively for the relief of the victims, and in the end, only unrelenting protest and demands for respect for human rights by the victims proved effective.

If primary industry was a victim of the heavy and chemical industries in the period of recovery, it was the creator of victims in the period of rapid economic growth. Heavy and constant applications of chemical fertilizers polluted the soil and water, and agricultural chemicals made the users both victim and the source of pollution. In addition, farming based on mechanization and chemical fertilizers caused a rapid decrease in the fertility of the soil, which in turn required more fertilizer to make up for the loss; in sum, a vicious cycle that prompted many people to forecast gloomy times for the agriculture industry and for the food economy of Japan, which was already dangerously far from being self-sufficient.

Some scholars look to genetic engineering and say that new fertilizer-free crop varieties may and should be mass-produced. Not all people, however, are optimistic about attempts to solve agricultural and ecological problems through engineering. Indeed, the pollution problems have made people increasingly sceptical about the nineteenth-century notion that what is born of science and technology can be remedied by new science and technology: the problems have, in fact, made scientists and technologists even less self-confident. Although many scientists, especially those in the United States, are unwilling to recognize ecology as a science on the grounds that it lacks objectivity and cannot be quantified, it is now the object of a great deal of attention. Science and technology began to be openly questioned in the 1970s, and the century-old philosophy of modern science is now being critically re-examined.

Keeping this in mind, let us refer to tables 1 and 2, which present the Japanese government's view of future prospects for Japanese technological development in comparison with the industrialized West. Table 2 includes findings from a survey of Japan's neighbouring countries in regard to technological development. From table 2 it is apparent that the Asian countries have developed their light, labour-intensive industries at an extremely rapid pace. Although for the time being these industries can be supported by domestic demand, eventually they must depend on export markets for their products, and their international competitiveness will greatly depend on an acquisition of high-level skills.

The long time needed traditionally to acquire skills is now being remarkably shortened by the introduction of new and efficient machinery. Industries whose raw material requirements are met domestically can remain internationally competitive through the introduction of new machines and technology. On the other hand, labour-intensive industries that depend on imported raw materials will quickly lose their international positions. As the introduced technologies become obsolete, the value of technology will come to depend on the locations of either resources or markets, and the advantage of cheap labour might be lost. Thus, it is very likely that developing countries will need to creatively reorganize their markets.

Table 1. Levels of technology and potential for its development in Japanese industries in comparison with the United States and Europe (1978: actual; 1985, 1990: estimates) (Japan is: ◎ = Higher; ○ = Comparable; △ = Lower)

Industry	1978 Level		1978 Development potential		1985 Level		1985 Development potential		1990 Level		1990 Development potential	
	US	Europe	US	Europe	US	Europe	US	Europe	US	Europe	US	Europe
Synthetic fibre	○	○	△	○	○	○	○	○	—	—	—	—
Textiles Spinning	○	○	○	○	○	○	—	—	—	—	—	—
Weaving	△	△	△	△	△	△	△	△	—	—	—	—
Apparel	△	△	△	△	○	○	○	○	—	—	—	—
General printing	○	○	○	○	○	○	○	○	◎	◎	○	○
Paper, pulp	△	○	△	○	△	○	△	○	—	—	—	—
Cement	○	○	○	○	○	○	○	○	○	○	○	○
Packing	△	△	○	○	○	○	○	○	○	○	○	○
Daily necessities	○	○	○	○	○	○	○	○	○	○	○	○
Flat glass	○	○	○	○	○	○	○	○	○	○	○	○
Fine ceramics (electro-magnetic, biochemical, optional functions)	○	○	○	○	△	○	△	○	○	○	△	○
Fine ceramics (mechanical, thermal, chemical functions)	△	○	△	○	△	○	△	○	○	○	○	○
Chemical Chemical fertilizers	○	○	○	○	○	○	○	○	△	○	△	○
Petrochemicals	○	○	○	○	○	○	○	○	○	○	○	○
Fine chemicals	○	○	△	△	○	○	○	○	○	○	○	○

Table 1. (continued)

Industry	1978				1985				1990			
	Level		Development potential		Level		Development potential		Level		Development potential	
	US	Europe	US	Europe	US	Europe	US	Europe	US	Europe	US	Europe
Light alloy rolling	△	○	△	○	○	○	○	○	○	○	○	○
Aluminium refining	○	○	△	○	○	○	○	○	|	|	|	|
Non-ferrous metals	○	○	○	○	○	○	○	○	○	○	○	○
Ferro-alloys	○	○	○	○	○	○	○	○	○	○	○	○
Iron and steel — Pig manufacturing	◎	○	○	○	◎	◎	○	○	◎	○	◎	◎
Steel manufacturing	◎	◎	○	○	◎	◎	○	○	|	|	|	|
Rolling	◎	◎	|	|	◎	◎	○	○	○	○	○	○
Special steel	○	○	○	○	○	○	○	○	○	○	○	○
Surface-treated steel plate	○	|	○	|	○	|	○	|	○	|	○	|
Steel pipe	○	○	○	○	○	○	○	○	○	|	|	|
Oil refining	○	○	△	○	○	○	△	○	◎	◎	○	○
Coal — Production	△	○	△	○	|	|	|	|	|	|	|	|
Use	○	○	△	○	|	|	|	|	|	|	|	|
Crude oil, natural gas	△	|	△	○	○	○	○	○	|	|	|	|
Non-metal, non-ferrous mining	|	△	△	|	|	|	△	|	|	|	△	|
Gravel	|	|	|	|	|	|	|	|	|	|	|	|
Thermal-power generation	○	○	△	○	○	○	○	○	○	○	○	○

○ ○   ◁ | ○ | ○ ○ | | | ○ ○ ○ ○ | ○   ○

○ ○ ○ ◁ ○   ◁ | ○ ○ | ○ | ○ ○ ◁ ○   ○   ○   ○

○ ○ | ◁ | ○   ◁ ○ ○ | | | ○ ○ ○ ○ | ○   ○

○ ○ ○ ◁ ○   ◁   ○ ○ ○ | ○ | ○ ○ ○ ○   ○   ○   ○

○ ○ | ◁ | ○   ◁ ○ ○ ○ | ○ ○ ○ ○ ○ | ○   ○

○ ○ ○ ◁ ○   ◁   ○ ○ ○ ◁ ○   ○ ○ ○ ◁ ○ | ◁   ◁

○ ○ | ◁ | ○   ○ ○ ○ ○ | ○ ○ ○ ○ ○ | ○   ○

○ ○ ○ ◁ ○   ◁   ○ ○ ○ ◁ ○   ○ ○ ○ ○ ○   ○   ○   ◁

○ ○ | ○ | ◁   ○ ○ ○ ○ | ○ ◁ ○ ○ ○ | ◁   ○

○ ○ ○ ◁ ◁   ◁   ○ ○ ○ ◁ ○   ◁ ◁ ○ ◁ ○   ◁   ◁   ◁

○ ○ | ○ | ◁   ◁ ○ ○ ○ ○   ○ ○ ○ ○ ○ | ○   ○

○ ○ ○ ○ ○   ◁   ○ ○ ○ ◁ ○   ○ ○ ○ ○ ○   ◁   ◁   ◁

Electricity Hydraulic-power generation
Transmission

Gas

Chemicals
Large showcases
Food processing (meat)
Food processing (cereals)
Packing
Metal working
Business
Printing

General machinery Forging and compressing
Sectioned materials
Textiles
Construction
Agricultural
Freezing and air conditioning
Export and engineering of plants
Atomic energy (light water reactor)

Table 1. (continued)

Industry	1978 Level US	1978 Level Europe	1978 Development potential US	1978 Development potential Europe	1985 Level US	1985 Level Europe	1985 Development potential US	1985 Development potential Europe	1990 Level US	1990 Level Europe	1990 Development potential US	1990 Development potential Europe
Electrical equipment (for medical use)	△	○	△	○	○	○	△	○	○	○	○	○
Electrical guages	△	○	△	○	△	○	△	○	△	○	△	○
Electrical materials	○	○	○	○	○	○	○	○	○	○	○	○
Semiconductors and ICs	△	○	△	○	○	○	○	○	○	○	○	○
General electronic parts	△	○	△	○	△	○	△	○	△	○	△	○
Electric machinery Household electric appliances	○	○	○	○	○	○	○	○	○	○	○	○
Microcomputers	○	○	○	○	○	○	○	○	○	○	○	○
Information processing (software)	△	|	|	|	○	|	|	|	○	|	|	|
Power generators	○	○	△	○	○	○	○	○	○	○	○	○
Lasers	△	○	△	○	△	○	△	○	○	○	○	○
Data bases	△	○	△	○	△	○	△	○	|	|	|	|
Computers	△	○	△	○	△	○	○	○	○	○	○	○
Aeroplanes	△	△	△	△	△	△	△	△	△	○	△	○
Automobiles	○	○	△	○	○	○	△	○	|	|	|	|
Cameras and other optical appliances	○	○	○	○	○	○	○	○	○	○	○	○

Ocean development	△	○	△	△	—	—	—	—	—	—	—	—	—
Housing	○	○	○	○	—	—	—	—	—	—	—	—	—
Atomic-energy industry	○	△	△	○	—	—	—	—	—	—	—	—	—
Other Social system (medical information system)	△	○	○	○	△	○	○	○	—	—	—	—	—
Social system (audio-visual daily information system)	△	○	○	○	△	○	○	○	—	—	—	—	—

Source: Institute of Industrial Technology, Ministry of Trade and Industry, *Sōzō-teki gijutsu rikkoku o mezashite* (Toward self-reliance in technology) (Tokyo, Government Printer, 1981), pp. 72–79.

Table 2. Levels of technological development of Asian countries (1978)

Manufactured good	Thailand	Indochina	Philippines	Malaysia	Singapore	Hong Kong	Taiwan	R. of Korea
Atomic energy equipment	1	1	1	1	1	1	1	2
Washing machines	1	1	1	1	2	3	4	4
Refrigerators	2	1	2	2	3	3	3	4
Lighting equipment	2	2	2	2	2	3	4	4
Communications equipment	1	1	1	1	1	1	2	2
Radios	3	3	3	3	3	4	4	4
Televisions	2	2	2	2	3	3	4	4
Computers	1	1	1	1	1	1	1	1
Electrical instruments	1	1	1	1	2	2	2	2
Resistors, condensers	1	1	1	1	2	2	3	3
Semiconductors	1	1	1	1	2	3	3	3
Batteries	3	3	3	3	—	4	3	4
Cars	1	1	1	1	1	1	1	3
Buses, trucks	1	1	2	1	1	1	2	3
Car parts	2	1	2	1	1	1	2	3
Motor cycles	1	1	1	1	—	—	3	—
Bicycles	1	—	—	—	—	—	3	3
Railway cars	1	1	1	1	1	1	3	3
Shipbuilding	1	1	1	1	3	1	3	3
Aeroplanes	—	—	—	—	—	—	1	1
Cameras	2	2	2	2	2	3	3	2
Boilers	1	1	1	1	1	1	1	2
Power shovels	1	1	1	1	1	1	2	2
Valves	2	2	2	2	2	1	3	3
Tanks	1	1	2	2	2	3	3	3

Bearings	—	—	—	—	—	1
Pumps	2	2	2	2	2	3
Waste-water disposal equipment	1	2	2	3	3	3
Agricultural machinery	2	2	2	2	3	3
Lathes	1	1	1	1	2	2
Textile machinery	1	1	1	1	2	3
Sewing machines (for home use)	1	1	1	1	2	2
Desk calculators	—	—	2	3	3	3
Electronic registers	1	1	1	2	2	3
Integrating watt meters	—	—	—	—	3	3
Wrist-watches	1	1	1	2	2	2
Lighters	—	—	2	—	—	—
Generators	1	1	1	1	2	3
Motors	1	1	1	2	3	3
Transformers	1	1	1	2	3	3

Notes: 1a. Figures 1, 2, 3, and 4 denote, respectively, that it would take "more than 10 years to catch up with Japan," "5–10 years," "less than 5 years," and "already comparable with Japan."
b. Enterprises covered here include those of native capital and joint ventures with advanced industrialized countries. For those items of which appraisals varied widely, figures are based on the majority of answers received.
c. Sixty-five Japanese manufacturers operating in Asian countries were interviewed about the 40 items shown in this table.
2. I would make no comment on this survey except to point out that it was conducted before the second oil crisis. The situation has changed, especially after 1985, because of the decline of the US dollar. Technology transfers were carried out in these countries on an unprecedented scale, which pushed up their position in the world.
Source: Nihon Keizai Shimbun Sha, Asu no raibaru—Oiageru Ajia no kikai kōgyō (Tomorrow's rival: The Asian machine manufacturing industry in pursuit) (Tokyo, Nihon Keizai Shimbun Sha, 1978).

Since technological innovation usually reduces employment, it is important that new markets be developed to absorb increased productivity. Second, the labour saved through innovation should be absorbed in the same branch of technology, which would require a new investment capacity.

A situation where investment is made for technological innovation and there is still capacity to invest is a typical picture of prosperity, a phase in which each investment calls for another. This situation, rarely seen, was experienced by Japan only in the period of rapid economic growth.

Such prosperity brings on inflation, which widens the gaps in the rates of growth between enterprises and industries. Gaps of this kind can pave the way for technological and managerial dualism, even on an international scope. In countries where social integration is not sufficiently high and a national consensus on the goals and means of development are lacking, political and social disorder and unrest may arise, which might paralyse technology and even bring on the loss of capital and technology. Consequently, countries responsible for their own development should be prepared to proceed carefully with technological innovation and should seek effective international co-operation.

Technology Transfer in Post-war Japan

I have dwelt upon aspects of the technological history of post-war Japan because my impressions, after having spoken with people from both developing and developed countries, lead me to believe that there is a great deal of interest in Japan's post-war technological, especially high-technological, development. There is a misunderstanding, however, that Japanese attainments in high technology have been due solely to technology transfer from the advanced countries. Anyone acquainted with the history of technology should be aware that this cannot be true, but not all people want this acquaintance, and further, people tend to expect too much of technology, without first learning exactly what technology is, especially what its internal logic or internal mechanisms are.

Technological change in post-war Japan has been remarkable, undergoing a radical change nearly every decade. Machines were renewed within a period of 10 years, and within the next 10, factory layouts were also changed completely. Some say that this is unprecedented. If so, it should be encouraging for those now struggling to develop their countries, because obviously it is not unattainable.

Of course, Japan did carry out a number of technology transfers from abroad, but many other countries have also had such possibilities. The first question we might ask, then, is, Why was it that some did not try to introduce advanced foreign technologies? And if they did try, but the technologies failed, why did they fail? By way of response to these questions, let us return to the Japanese case.

We may first of all point to Japan's need to recover from its heavy war damage, an urgent need obvious to anyone. But not all industries were successful in introducing new technologies, nor earnestly willing to do so.

To take the most successful case, the steel industry chose to introduce rolling technology, the final manufacturing process. That was logical, since at the time rolling was where Japan lagged farthest behind the other industrial nations. Any other country in a similar situation would have done the same, and in fact that is what some South-East Asian countries are now doing. This kind of technology transfer makes it possible to economize in construction, operation, and fuel costs (the Japanese steel industry economized on construction costs by 30 per cent) compared with technology requiring new construction, such as that for sintering, blast-furnace, or converter processes in the continuous operation of iron and steel production.

When the LD oxygen furnace was introduced, for example, it represented an optimal choice, making full use of the advantages of the late comer. In 1979, the ratio of LD converters to total furnaces was 81 per cent for Japan, compared with 66 per cent for West Germany, 56 per cent for the United States, and 21.4 per cent for the Soviet Union. As for blast-furnaces with a capacity exceeding 3,000 m^3, in 1985 Japan had 12, the Soviet Union 2, West Germany 1, and the United States none.

Japan began using LD converters in the 1960s. The US steel industry had built many open-hearth furnaces in the 1950s, which made it unnecessary and untimely to switch over to the LD converter. Besides, because American iron-ores are highly sulfuric, the open-hearth furnace is better at making high-quality steel.

At the time, most LD converters were operating at an average rate of 40 charges per day, and Japan's Yawata was running them with 50 tons per charge; the corresponding figures for an open-hearth furnace were 5 or 6 charges per day at 200 tons per charge. The productive capacity gap (2,000 tons versus 1,000–1,200 tons per day) was obvious. Moreover, the former had the advantage in fuel costs (6 to 10) and in construction costs (1 to 2).

According to a view dominant in Japan, theoretical innovation in this branch of technology did not take place globally in the 1960s, and, consequently, competitiveness naturally depended on a factory's scale of operation. In Japan at the time, there were few obstacles in the way of enlarging the scale. And, as the scale grew larger, the Japanese steel industry proved to have advantages over others in the operational skills needed to handle the growing capacity. Thus, Japan turned out to be the rare case in which the late starter had the advantage.

The Japanese industry's own efforts toward innovative design of the entire manufacturing process, from factory layout to factory location, were another contributing factor. Thus one can see that an accumulation of modifications in operation and processing or both, minor as they might appear from the point of view of engineering, can have effects that are not at all minor. It also becomes evident how standardized the technologies were in this industrial branch and how much steel technology had matured. Factors such as operational skills and factory layout, what may be called soft technology or technology management, can greatly increase competitiveness in this branch of industry. Strangely enough, however, little attention has been paid to this fact. There seems to be an unfortunate tendency for engineering-oriented technologists to neglect the question of the extent of maturity of the relevant technologies when discussing the ability to develop technology.

As for the failure of technologies to transfer successfully, the second of our two questions posed above, it should be pointed out that modern technologies differ from pre-modern technologies in that they are freely transferable on a commercial basis. However, they cannot be freely joined; they require related technologies and the availability of supporting services. So, when a technology fails to transfer successfully despite the enthusiasm and serious efforts of the parties concerned, or when a technology, once transferred, does not meet expectations, most likely it is because it lacks the necessary pre-conditions and supporting services. In other words, the cause of failure can be found in inadequate feasibility studies concerning optimal type, level, and scale of transferable technology or in a lack of efforts in preparing the necessary conditions for the transfer, such as enhancing the levels of fringe technologies and services.

Japan succeeded because it already had the right conditions; the related technologies had been sufficiently readied to make the transferred technology operative and thus it could further develop the transferred technologies.

This being the case, it is perhaps worthwhile to look back on Japan's history of technology to find out how those pre-conditions and supporting fringe technologies were built, because they were not brought into being on short notice after the war, nor were they provided by foreign countries. Indeed, Japan won its success at tremendous social cost. The problem of occupational diseases was and is even now very serious. In addition to the problems of pollution and such phenomena as *mura-tsubure*, there is another problem perhaps worth mentioning.

In the old-style ironworks, rolling was a process that required skills that could be acquired only through many years of experience. Half-naked, muscular workers, with sweat glistening on their skin, would toil long, and yet in a high-spirited manner, creating a sight that would make any observer stand in awe of human labour. This way of work has already passed into history.

Yet, it later turned out that the latest computerized factories had to consult the old skilled workers for advice; despite automation, their skills and knowledge were still needed, as automation demanded highly and multi-skilled workers. Automation may have displaced the value of single-craft skills, but the latest automated processes often reveal themselves to be inferior to the human skills of former days, as will be discussed later. Suffice it to say that the skills of management and workers constituted important factors in the success of Japan's technology transfer.

The Japanese steel industry, with its highly matured technology, is now being pressed to choose for its survival among three alternatives: (1) to remain as a supplier of materials, not only iron but also new materials; (2) to change over to a compound-processing technology that ensures higher value added; or (3) to try to survive in a new general engineering field. The Japan Steel Corporation, for example, while emphasizing the development of new materials and electronics, continues to develop new iron-manufacturing technology.

The equipment of the Japanese steel industry is now nearing the last phase of its life in an economic sense, but capital investment for its renewal has not been active. Because the location merit of technology tends to shift globally from consumption to resources, developing countries are planning to build their own steel mills. For an output of 1,000 tons per annum, 1,000 tons of steel will be required for construction and equipment, which would mean that demand for steel would be on the rise for a long time to come.

However, it is now down, which compels the existing steel manufacturers to cut back in their operations. For the cut-back not to cause a great income loss, it becomes necessary for the manufacturers to develop and invest in higher technologies different from those of the production expansion period, and this acts as a brake on capital investment for equipment renewal. Steel technology is now said to be in its last stage of glory. Whether or not this is true, we may be sure that steel technology is no longer the leading technology.

2

The Japanese Experience: The Problems and Attempted Solutions

1: Expectations from Outside

The United Nations University, in commissioning this project on the Japanese experience, observed that Japan, once an importer of modern technologies from the West, has now developed itself to the point of being an exporter of the latest technologies, and, further, that what it was that made such a transformation possible was a matter of great interest to developing countries.

At about the same time, a meeting of experts sponsored by the University released a document that concluded that theories of development were in a state of "disarray." Although it did not state explicitly reasons for the disarray, I would like to present some of my thoughts on the basis of experiences I had in sessions I attended at that function. First, developments invalidate theory; second, theories have long been questionable, but persuasive data to overthrow them have not been available; and third, though development needs have become diverse, theory has failed to keep up.

For these reasons, Japan has become the focus of attention among the developing countries, and I find this interest perhaps may have great practical value and, at the same time, it poses an exciting challenge: (1) Might the Japanese experience not be made useful in filling the information gap caused by the disarray of development theories? (2) Could it not provide practical suggestions to meet development needs?

Though their interests were too diverse for easy generalization, one thing seemed certain: too much attention was focused on Japan's "miraculous" post-war technological success; little interest was shown in the social and historical context in which transferred and domestic technology were able to flourish. Moreover, development in general tended to be discussed with technology in general, with no awareness of the particular and multifaceted relationship between them.

Technology can be discussed practically only on the basis of concrete data and for the purpose of discovering possible solutions to existing problems. Any discussion of technology that is too generalized tends to veer away from technology as such and slide into abstract arguments on policy and the international politics of technology transfer.

Such discussion may reflect their own national experiences, which are none too pleasant. Where technology transfer has already been more or less institutionalized through the ODA, the politics of technology may have proved both important and inevitable. In many of the aid-recipient countries, academic freedom to carry out scientific study of technology and development, which may reveal weaknesses and conflict in development policy, is restricted. Some intellectuals in these countries are calling for a moratorium on development.

Under these circumstances, the more urgent the development needs, the greater may be the danger of a political treatment of the problem and of neglecting to be alert to the inner mechanism of technology. In other words, by politicizing a problem that requires a technological solution, one might well be producing a result more harmful than beneficial.

As for the relationship between development and technology, just as there are different levels of development, ranging from the level of village or province to that of an entire country, even to the international level, so there are different levels of technology. This is an important fact to be considered when transferring technologies for the purpose of national development.

If technology is a means to development, it is undesirable that the end and the means exist in a one-to-one relation; it is preferable to have several means. Yet, a mature technology requires materials, processes, and equipment that are all standardized, and the methods, equipment, and technologies used in the manufacturing operations are also fixed. Thus, in the case of a transfer of a mature technology, the recipient country has little possibility of adapting or improving the technology. Here, then, the relation between the end (development) and its means (technology) must, in practice, be one to one. In other words, the end is restricted by its means.

That is where the problem of technology begins to emerge in a developing society. In order for a country to avoid the diseconomy of a technology transfer that would result from introducing a mature technology, it is important to (1) select a technology with a rich potential for expanding and upgrading the links among existing sectors of industry and (2) observe and gradually adjust the transferred technology to successfully meet the needs of country, region, and local community, both in quality and quantity. Technology modified in this manner to meet local conditions and needs is what we refer to as an intermediated technology, an alternative technology free from standardized pre-conditions and processes for operation.

In general, modern technologies are freely transferable but not always easily integrated. Therefore, the range of choice of technologies for transfer is usually limited. The possibility for a developing country to find a ready-to-transfer alternative or high technology suited to its own conditions is quite limited. But a developing country often has no other way to solve its technology problems, hence the necessity for an extensive knowledge of technological science and technological history.

The view one often encounters of the general relationship between development and technology is too optimistic; it tends to exaggerate the diachronic and cross-cultural nature of technology and also to expect too much of technology's power of breakthrough. The impression is that the problem of development and technology has not been rightly grasped in all its implications. This seems also to be reflected in the way in which alternative technologies and intermediate technologies are frequently discussed. I believe a real understanding of the painful experiences of developing countries in the past 30 or so years is still not evident. There are gaps between expectations and realities and between wishes and capabilities with respect to development and technology, and therefore it is urgent a solution be found.

Because of the complexity of today's development problems, the information the Japanese experience might provide seems inadequate to meet the needs of developing countries. But because it is the experience of a late starter, it should enrich their knowledge.

As stated earlier, we will avoid here the politics of technology transfer as much as possible. The approach here is one free of non-technological values, and our aim is to present practical results from case-studies and avoid speculative and abstract reasoning. It is important to find opportunities for generalization, not the other way around.

Our hope is for a dialogue based on actual cases of technology transfer; it may not be possible to achieve all our expectations, as gaps in perception of the problem are bound to occur from country to country. But continual and multi-pronged efforts must be made to bridge these gaps.

Although the social sciences have working methods not bound by national borders, quite obviously, each social scientist has his or her own nationality. The way in which one investigates, constructs, and pursues work on a problem is to some degree rooted in the characteristics of one's native country. Therefore, in treating the theoretical handling of our problem, we must first determine what is common to the many diverse national characteristics.

Until now, however, efforts to do so do not seem to have met with success, at least not in the way we had expected. The present volume thus represents an attempt to cover the distance that separates us from this goal, a step that may be termed a methodological prelude to better dialogue. It is one of many possible methods that may be used to attack the problem and, consequently, is subject to correction and change. In any event, we do not advocate methodological exclusivity. Likewise, we do not wish to make the Japanese experience a dominant or universal model; but neither do we intend to minimize or ignore it.

Our methodological pluralism, therefore, does not exclude cross-national or diachronic analysis; both are necessary. But their descriptive powers concerning technological problems, even if not yet exhausted, have become increasingly unable to meet real needs and expectations. New problems that can no longer be handled by earlier social science theories and new propositions have come to assume importance. This is why the Japanese experience seems to have attracted so much attention abroad as a case for study.

It should be understood that the social sciences, necessary and useful as they are, are not almighty. They can breed error, just as other sciences can. Unless it is realized that their validity is circumscribed by time and circumstance, whatever is scientific in them will be lost. The social sciences as discussed here denote an intellectual discipline based on the theorization of confirmed contemporary and historical reality and capable of making policy proposals.

Development problems can and should be solved finally and decisively only by each nation concerned. All we can do is to help other nations in their attempts to develop. Whether the Japanese experience is worth studying must be judged by third-world intellectuals themselves, and different nations may very well make different evaluations as a result of their different, often unique, development problems.

The results expected from a study of the Japanese experience tend to be exaggerated, and as knowledge of the experience accumulates, interest tends to diversify and demands to grow.

A gap may arise between what we can do to meet the demand for information and how much we want to release. Too little information may generate a too keen and unbalanced interest. A most important part of our effort, therefore, should be to prevent this from happening. It may be very difficult to acquire an accurate and deep understanding of another country, but the person who attains it will learn to know his own country better.

Our approach to the Japanese experience in technology and development began by classifying development problems on the basis of our own group's experiences in research tours to developing countries, in gathering scientific information, and, more important, in continued dialogue with intellectuals from these countries. We tried to elucidate how those problems were overcome in Japan or why some remain to be overcome. In a sense, our approach to development problems is provisional.

2: Japan's Response

Economy and Technology

Two views have characterized Japanese studies of the relationship between development and technology in Japan: (1) technology is a dependent variable of the economy; (2) technology possesses its own inner mechanisms that make it relatively independent of non-technological elements. I would like to expound the first view. As mentioned, in the middle of the nineteenth century, Japan was forced to open its doors to foreign trade and to establish diplomatic relations with the Western powers, which had already achieved their own industrial revolutions. Compelled to accept unequal treaties, Japan was on the brink of being colonized. The Tokugawa regime, however, lacked the leadership necessary to surmount the crisis and preserve national independence. So it was up to the new Meiji government to build a strong modern state, through enhancement of the country's wealth and military power and through a series of political and social reforms. A popular slogan of the time was *Datsu-A Nyu-O*—"Withdrawal from Asia and Entry into Europe."

The Meiji government introduced modern technologies and spurred the nation's economic development, with the aim of building a capitalist economy at the hands of the state. The emphasis in industrialization was placed on a realization of self-reliance in weapons and arms supply, deemed necessary for both national defence and controlling domestic discontent. Because foreign currency was necessary to finance industrialization, the government established state-operated industries and factories, into which it introduced modern technology.

Although Japanese industrial technology was largely pre-modern in character until the end of the nineteenth century, a few state-operated factories were exceptionally equipped with imported modern technology and machinery, operating, however, to meet the needs of government, not of the general public. Economic rationality was belittled because domestically produced goods were higher in price and lower in quality and durability than imported goods. The state-run, bureaucratically managed factories proved unprofitable and were sold to the private sector. Because the change of hands involved not only equipment but also engineers and workers, the result was a large-scale secondary transfer of technology.

The military arsenals, which remained under state control, were better equipped and in possession of higher levels of technology compared to the other factories. The favouritism was also evident in the application of energy: the military had steam-power, while the private sector generally had access to only human power or water-power.

Thus, industrialization and the national formation of a modern technology network were brought to completion in Japan on the basis of a structural dualism—government vs. private; industry vs. agriculture; big enterprises vs. small; heavy industry vs. light industry and handicrafts; and central vs. provincial.

The view that technology is a dependent variable of the economy thus stresses the necessity of the modernization of Japan, for, without a modern economic system, technology could not have developed. It does not aver that the role of the government was of exclusive importance, because there was an active response from private interests. When, some 20 years after the Meiji Restoration, government enterprises were sold to the private sector, technology transfer ignited a great entrepreneurial boom. This centred on light industries, notably textiles and food processing, but paralleled or preceded the development of mines and railroads.

The Meiji government, nationalistic towards other countries, took an autocratic, "statist" position toward its own people. So what should have been regarded primarily as the economic evils of capitalism tended to be viewed as political evils, and a chain of urban riots and peasant revolts resulted. It is worth noting, however, that antistatism did not necessarily mean a denial of nationalism among the Meiji Japanese.

Thus, Japanese industrialization, even if a capitalist development "from above," followed the historical path of light industry first, then heavy industry. However, the initial dualism between government and private enterprise in favour of the former remained in Japan's industrial structure. It favoured big business, which took over the government enterprises, and the gap between the big and the small came to be fixed not only in technology but also in the ability to develop technology.

Since the costs of a technology itself, and also of its development, are high, technology is subject to economic laws. Those who are late to enter a business are exposed to competition from those advanced in technology and in possession of the ability to develop it. Thus confronted both within and outside the country and in need of keeping abreast of technological innovation, the government and large private enterprises often separate parts of their technologies and manufacturing processes into new, independent companies to disperse risk and lessen fixed costs. Skilled workers also often become independent when business conditions are good, acting as producers and suppliers of goods required by the parent enterprises.

The diversity of smaller enterprises and their high technological standards are considered to have been the basis of Japan's strength in technology. Although some of the smaller enterprises failed to keep abreast of the parent companies in technology development and had to drop out from the ranks of subsidiaries or subcontractors, those that managed to secure a high-enough technological standard and the ability to enhance it were able to expand their transactions and stabilize their positions.

From the standpoint of the parent companies, this separation of processes enabled technological spin-offs that acted as buffers and a reorganization of manufacturing processes according to the logic of capital. In Japan, this kind of relationship has long been viewed critically as the source of wage dualism between larger and smaller enterprises and of the problem of the smaller being tyrannically dominated or exploited by the larger.

Although spin-offs have sometimes produced large, technologically advanced conglomerates that eventually succeeded in establishing world-class concerns, very many have, to the contrary, caused the parent company to fail in accumulating sufficient technological power when it was badly needed, resulting in bankruptcy. Both results have occurred in mining, which seems to indicate that mining demands two things to modernize: a complex system of technology and a careful study of how management should relate to that system.

Some economists maintain that dualism in management and in technology weakened during the period of rapid economic growth in post-war Japan. But I believe that dualism is evident in big factories even today; it can be found in the labour structure, that is, obvious gaps exist between the jobs, skills, wages, and welfare benefits of regular workers and those who are subcontracted or part-time. The latter do not have permanent employment, which results in a high percentage of job changing. The rapid progress of technological innovation and changes in the leading sectors have made it difficult to organize national unions in Japan. Labour unions have been over-whelmingly enterprise unions or in-house unions of mixed lots of workers.

In the sectors of technology that had achieved global maturity before 1960 and that have seen little change since, innovation has tended to take place only in such directions as enlarged scale and capacity of operation, increased speed of operation, and expanded automation in pursuit of merit of scale. This is particularly so where the technology has already been standardized. But in some areas, machine tools, for example, the ability to accumulate superskills and to develop technology is increasingly to be found in the smaller-scale enterprises. This is especially true in some of the most advanced technologies.

The dual structure, which an analysis from the first viewpoint would say characterizes the Japanese type of industrialization, thus may be said to per-sist even today, though its forms and dimensions have changed. However, the problem of dual structure in industry in the developing countries, subject as it is to both domestic and international industrial structures, must be treated as being qualitatively different from the problem in Japan.

A period of dualism may be inevitable for a country late in starting indus-trialization. When I accompanied a group of scholars from developing coun-tries visiting factories in Japan, I noted they were impressed that the dual structure system encouraged competitive coexistence of the parent and sub-contracting enterprises, not merely coexistence with little mutual contact. Further, at a factory making products for export, even though its scale was much smaller and the equipment used much older than at factories in the visitors' home countries, they saw immediately that good operational skills and high managerial capability more than offset any such disadvantages.

This was their "discovery of Japan." It should be kept in mind, however, that the Japanese approach—or the concerns prevailing among Japanese academics—cannot be applied to the developing countries without adjust-ments or revisions in the light of the existing conditions of these countries. What the Japanese take for granted is not always understood or accepted by other nations. This is due as much to differences in natural conditions, re-source allocations, and production activities as it is to historical and cultural backgrounds, and the differences should not be reduced simply to develop-ment stages.

So, whether analyses based on the first view are applicable in treating the development problems of developing economies—and to what extent—must be re-examined through in-depth case-studies in each country and in each industry. This task can be accomplished only through international collabora-tion. Our study of the Japanese experience, therefore, for its conclusions and analysis to serve a useful purpose, must be supplemented. Though our work has benefited from co-operation with other countries, it has only just begun.

The Fixed Logic in Technology

The second view, that technology is independent of non-technological elements by virtue of its inner mechanisms, addresses directly the severe difficulties of technology transfer. This contrasts with the first view, which regards technology transfer as a natural historical process. While the first view addresses the aftermath of transfer, the second considers how the transfer begins and progresses and details of the practical and functional problems of technology at the shop-floor level. These aspects are supposed nearly independent of non-technological problems, of political and economic laws and customs.

Thus, the second position assumes that the inner mechanisms of technology, the laws firmly set in technology, cannot be arbitrarily changed or modified. Thus, when a technology is transferred with some political or other intention, this view facilitates the clarification of why it did or did not succeed, i.e., whether for technological or non-technological reasons.

For example, if a transferred technology encountered trouble after initial success, working on the basis of this second view, one would pin-point the trouble in raw materials, poor operation, maintenance and repairs, improper management or technology control, or in the general plan itself.

For example, before the Meiji Restoration, a commercial representative from the Netherlands, then the only Western country allowed access to Japan, stated in a secret report that the Saga clan in Kyushu was attempting to manufacture a steam-engine:

The Japanese seem to simply assume that they can master this manufacturing technology, but the only equipment they have are poor furnaces and moulding factories. Iron of low quality is processed with poor machines and by unskilled workmen. They have a will to manufacture, but little means for it.

This brief report gives an idea of the technological picture of Japan at that time. Modelling their endeavour after a finished engine they had seen, members of the Saga clan set about to do nearly the impossible with the help only of a technical manual, but their supply of fire-brick was insufficient (because the Saga area had no natural resources), and they knew little about what the interior of a furnace should be and what temperatures were required. Further, they had little knowledge of what quality of iron to make and how to process it. What they did have was an immense desire; what they lacked was the means. There were too many obstacles: raw materials, fuel, instruments, machines, methods, to name those most prominent.

From what the Dutch representative observed (it remains unclear whether he was an engineer), it seems evident that a great gap in technology separated the West from Japan. Obstacles lay everywhere blocking Japanese efforts to adopt technology. Nevertheless, they remained convinced their goals could be achieved—no matter the problem—through mobilization of all their traditional skills and abilities. Opinion may vary as to whether that was a mindless or an admirable position.

The second view recognizes that some technologies were already present in Japan, though of low standards, and attaches importance to them as the Japanese predecessors of modern technologies, even if they were of little direct use as they were then, and had to be wholly redesigned; their very existence made a great difference in the future of Japan and its development of modern technology.

In those pre-Meiji times, Japan was divided into some 240 large and small feudal domains, all variously endowed. Thus, it was politically impossible to bring together the empirical knowledge and skills that had been accumulated by carpenters, masons, builders, forgers, potters, and other craftsmen of different domains into a national technology plan. Moreover, Western scientific knowledge was regarded as serious political criticism of the system, and its students were in danger of conviction and execution for treason.

The rush to import Western science and technology began only about 30 years before the Meiji Restoration, when engineering experts armed with modern science were free to appraise and implement the empirical knowledge and skills of their predecessors without fear of conviction for political offence. The foreign experts later hired by the new Meiji government could not be expected to appraise and use traditional Japanese knowledge and skills, so all they could do was introduce their own technology as it was, a point we will discuss later in more detail.

Such was true even in the successful transfer of spinning technology. A foreign engineer, whenever and wherever he may serve, tends to be a believer in the transcultural and diachronic validity of technology. That is where his usefulness and his limits will be found. Regarding iron manufacturing in Japan, only the Japanese engineers were able to domesticate the transferred foreign technology. And, as is well understood, technology can spread only after it has stabilized.

Scholars holding the second viewpoint insist there is no leap in technology. They say that rapid progress in technology, whether vertical or horizontal, can be achieved only after proper and adequate operational and manufacturing skills have been accumulated. They are interested in the process in which accumulated technologies and skills come to be applied. Technologies concerning materials, processing, and design are developed individually before being integrated, and only after this process are theoretical levels of engineering and technology refined, and thus the applicability of the technologies is assured. Adherents of the second position grasp this process of technological development as one peculiar to each country, to each time, and to each enterprise or plant.

More important, this position is attentive to the role of labour as a factor of technology, especially to the indispensability of engineers and skilled workers. It has often been assumed that production and productivity gaps between countries, regions, and factories would be narrowed or altogether removed if machines and equipment were standardized. The fact is, however, that conditions under which different factories operate vary widely, and the closing of the gaps may be far more difficult than imagined.

These differences should first of all be attributed to the human factor in technology. In the latter half of the nineteenth century, Japanese spinning factories adopted ring spinning, the most efficient spinning technology in the world at that time. The ring spinner was far easier to handle than the mule, and it was hoped that the new technology would raise production efficiency three to four times, as it had in Great Britain. In actuality, however, as an official report of 1903 declared, productivity in Japanese factories was as low as one-eighth of what it was in British factories, where even obsolete machines were in use.[9]

The report further describes the British spinning workers as professional soldiers and the Japanese as rabble, and attributes the Japanese weakness to the very short average terms of employment; most workers quit within two months of employment. Most obvious in the report is that, while in theory a farmer's labour is equivalent with that of an industrial worker, in reality, a farmer cannot easily make the transition from the soil to the factory.

A similar situation was evident in the watch and camera factories—where primarily originally agricultural labour was used—of some South-East Asian countries. In their first years of operation, the percentage of end products meeting standards was as low as a tenth what it was in Japan. What this shows is that farm skills and labour fall far short of the needs of industry and, not surprisingly, the work roles of factory workers and farmers are not easily exchanged. Because state-of-the-art machines require less skill from labour, the productivity gap between skilled and unskilled workers and between the new and old industrial nations has become narrower than when simpler machines and tools were in use. But the gap remains, and the apparent narrowing should not be misinterpreted; the latest machines, though efficient, are far more expensive than the ones they made obsolete. And throughout technology, the human factor, the skills and accumulated experience, are indispensable at any time, at any place, in any sector. As the machines and equipment change, so do the type and the substance of skills required to operate them.

In another example, there is a steel mill with the latest equipment which has an automated control centre that is notified of every activity in every process at the mill. If any process deviates from the programme, the centre is immediately informed. On one occasion, a process was found to be in trouble. The prescribed corrective measure was taken, but it had no effect and the trouble spread quickly downstream. Investigation determined that the other processes and the programme itself were functioning properly, and the trouble was found to be confined to the process that had shown misfunction.

Because the mill is wholly automated, the cause of a malfunction can usually be determined and remedied through an examination of records;

which, however, can be a two-week job. The search would mean suspended operations and perhaps customer demands for compensation for losses and delays. It could even mean having to remove some equipment for repairs.

So the mill in this case looked to veteran skilled workers for help, even to retirees. They were quickly brought in by chartered plane, and once there, closely checked the processes, paying close attention to sound, light, temperature, and the shapes of processed goods in order to determine the problems.[10]

Automated equipment is designed to fit the movements and judgements of skilled workers and will never favourably compare to human labour that is highly skilled and efficient. In a modern automated plant operated at a high degree of stability, the importance of human efficiency and skill may not be as great as it once was, but it remains a necessary ingredient of the manufacturing process, especially at the start-up of production and for maintenance checks. In the chemical industry, for instance, groups of skilled workers, now retired, have organized businesses to provide help when technology breaks down.

Needless to say, what is possible theoretically is not necessarily possible in practice. A manufacturing technology must be established under restrictions and conditions vastly different from those of an experiment, which usually can be stopped and resumed at any time. Science and technology differ greatly here, and thus a correct diagnosis will not necessarily lead to a cure.

3: Why Do We Begin with the Meiji Restoration?

The Sixty Years towards Self-Reliance in Technology

At the beginning of the present report, I presented my own thoughts on development and technology in post-war Japan, a theme not included in our project activities on the Japanese experience in technology transfer and development.

I included it because, during many of our discussions with collaborators from the developing countries, interest centred on that aspect of the Japanese experience.

But it must be kept in mind that the technological development in post-war Japan was possible only because of the nation's pre-war legacy of development in the technology network. This cannot be overemphasized because the favourable conditions for technology transfer did not exist only in Japan. In fact, Japan was unable to attain complete self-reliance in technology until after World War II, especially in the 1970s, but this was made possible only because it first progressed through the recovery of the pre-war level of technology development. The ability to absorb state-of-the-art foreign technology was ensured by the country's first regaining the pre-war levels in the technology-supporting sectors and services. This recovery was helped by technology transfers, but more important is that it took place along with demilitarization. This differentiates the formation of the post-war technology network from that of the pre-war period.

The isolation and set-backs technology suffered during World War II caused Japan to lose much of its ability to develop technology, and the country fell drastically behind in this area after the war. Even today, Japan has much in common with many developing countries. The only difference is in the level and scope of national technology formation. That is why I have placed Japan as a front runner of the technologically less-developed group.

And yet, the Japanese experience differs from that of the presently industrializing countries in the method and time of technology network formation. In particular, the time difference has had much to do with whether technology transfer will prove easy or difficult.

Thus, our study of the Japanese experience in forming a national technology network through technology transfers should attend to the different phases of transfers, which corresponded with the changes in the level of technology in Japan. Consequently, initial attention must focus on the time when Japan began to absorb foreign technology, the Meiji Restoration, because this time factor influenced both the direction and the pace of the network formation.

The Meiji Restoration represented a political turning point. Though it was not a turning point of technology, it did pave the way for one. Only after the Meiji Restoration were there suitable conditions in terms of politics and socio-economics to domesticate and develop imported technologies. In the earlier cases of technology transfer, the Tokugawa regime had failed to create these conditions.

By the same token, the turning point in Japanese technological development after World War II would never have been reached without a series of reforms carried out as a result of another great political change, namely, the nation's defeat in the war.

The Meiji Restoration and the defeat in the war both clearly illustrate the relationship between technology and political and social factors. However, the political and social conditions of the restoration greatly differed from those of the defeat; the restoration was far more decisive for technology than the war as a turning point. That is, the Meiji Restoration represented an attempt by an agrarian society to turn itself into an industrial society, whereas the post-war development meant a change in direction and an upgrading of levels in a society that was already basically industrial. The latter experience thus was not primary, and as a secondary experience it was less painful and shorter than the first.

Technology Transfers Accelerated by Self-Reliance

Social and cultural conflict in Japan was far more serious at the start of industrialization in the Meiji period than after World War II. While the two periods are both characterized by a blind worship of foreign technology, and both experienced a flood of technology transfers, they differ in the way they absorbed them. Regarding the post-war period, for example, the formation of a national technology had already been basically completed by the 1920s, and the technology transfer after World War II was completely in the hands of the private sector.

In the 1920s and again in the 1960s, the formation (and, in the 1960s, the recovery) of a national technology network did not lead to a rejection of foreign technology; rather, it made it easier to absorb higher-level foreign technologies; indeed, it accelerated the process.

In examining this historical background, let us first focus on the period from the Meiji Restoration (1868) to the 1920s. At this time, government involvement in technology and industry was relinquished in favour of business groups.

One thing the developing countries share with Japan is that, once having set forth on the road to industrialization, all have had to tackle social and technological problems common to once basically agrarian societies. On the other hand, some of today's developing countries could choose to reject industrialization, as some in fact have. This may be a commendable choice for some, but not for all. As for Japan, it resolved more than a century ago to abandon being an agrarian society, and the national consensus on this is worth special note. Although the nation had agreed on the transformation, there was no unanimity on how it should be accomplished. Even today there is debate regarding Japan's choice to industrialize. Obviously, however, Japan has gone too far to revert to being an agrarian society.

Nevertheless, rising agricultural productivity supported Japan's industrialization and its growing population. And now, agriculture has become increasingly dependent on industry for farm machines, fertilizer, and agricultural chemicals. Japanese agriculture today could not survive without industry. So the question arises as to whether countries that have chosen to remain agrarian can continue without facing insuperable difficulties.

This seems especially true for countries with rapidly increasing populations. Since the international environment when Japan struggled towards industrialization was quite different from that of today, the developing countries may never experience many of the difficulties and pains that confronted Japan, though they will likely face others. It is our hope in presenting the Japanese experience that they will learn whatever lessons might be helpful in steering them away from, or at least minimizing, those difficulties and pains.

In this study, particular attention has been paid to the view expressed by some of the participants in our project which says that a comprehensive study of the Japanese experience should begin with the Meiji Restoration as the primary experience of modernization in Japan, but that, in regard to technology, greater relevance (for the developing countries) is to be found in the period since the 1920s, when global technological monopolies came into being. We do not agree with this view, however, because, for one reason, monopolized technologies have always been the most advanced technologies, which are not always useful for developing countries. What is urgently needed now are intermediate or alternative technologies.

From Agrarian to Industrial Society

The transition from an agrarian to an industrial society undertaken beginning with the Meiji Restoration meant that farmers were now working in the manufacturing and service industries on a nation-wide scale. This transformation entailed a lifetime of effort in acquiring new skills and experiencing conditions that were entirely new.

In the initial stage of industrialization, farmers and workers can perhaps assume each other's tasks, but as industrialization progresses, the interchangeability of roles is gradually lost. A farmer can only hope to become an unskilled worker, and an industrial worker can only expect to perform well as a farm labourer, not as a farmer. For farmers, industrialization brought a process whereby they necessarily became principally farmers, agricultural specialists, no longer able to maintain sideline occupations. The change began with the Meiji Restoration and gradually spread throughout the country. Thus Japan became an industrial society, and it became impossible to return to what it had been.

During this period of social change, the role played by women from rural areas was great. In the textile-led transformation of Japan into an industrial society, females had begun to account for more than half the industrial labour force by around 1910. As light-industry development gradually gave way to the stage of heavy- and chemical-industry orientation, males began to exceed females in the labour market. Also, a little later in this period (late 1900s), more graduates of the imperial universities in Tokyo and Kyoto, who were expected to form an élite corps in the service of national interests, were choosing business careers in big *zaibatsu* corporations and banks rather than in the government bureaucracy.[11]

Nevertheless, juvenile female textile labourers, forced to work long hours under severe conditions, played a central role in Japan's development of self-reliance in technology. In families who had been squeezed out of their farm villages, the men's wages alone were not enough to support their families, and it was necessary for wives and children to earn what they could from odd jobs they could do at home. This phenomenon has been referred to as *zembu koyo* (whole-family employment), to be distinguished from full employment. This whole-family labour corresponded with the practice of young women labouring in the spinning mills, sending all their extremely low wages back to their home villages to support their parents.

This phenomenon and its related problems suggest the need to consider not only the economic aspects of technology transfer and development but also the social and historical changes that result. Thus, focusing on the technological development of Japan after World War II would not give an accurate and practical analysis of the Japanese experience. Such a study must begin with Meiji, when Japan was a late starter.

3

Theoretical Summary: A Preliminary Examination and an Interim Conclusion

Technology in Theory—The Five Ms

Although science and technology are closely interrelated, each is relatively autonomous, perhaps similar to civilization and culture. Where civilization is universal, culture is individual. Today we live in an industrial civilization, an age in which nations are often ranked according to the level of their techno-logical development. Such a narrow basis, however, overlooks cultural differ-ences among nations.

At the same time, cultural differences are often taken to mean the supe-riority or inferiority of one culture to another. It should be too obvious to mention that cultural differences are not equivalent to differences in value.

By the same token, neither the wooden weaving machine nor the one made of metal has an advantage in principle over the other. Their mecha-nisms are based on the same scientific principle, and in this sense they are equal in value. They differ in durability or efficiency, perhaps, but these and similar differences should not be assumed to indicate the superiority of one over the other.

Any evaluation of them from the perspective of national development should also consider, besides productivity, the relative advantage for each of manufacturing operationability, procurement, maintenance, and repair to the indigenous user. The relative advantage of the wooden or the metal loom will depend on the production purpose and on what sort of consumer need its products must satisfy. For example, one option may be mass production and mass consumption; or goods may need to be produced and consumed in small quantities. The particular needs and conditions will determine relative advantage.

No society or culture exists without a technology of its own. Similarly, no community or culture today can avoid contact with the outside world, with foreign technologies, a fact that may hold fortunate or unfortunate consequences. As pre-modern, pre-industrial societies and cultures confront problems of population increase, many have been made keenly aware of the limitations of their traditional technologies. Consequently, there is an eagerness to introduce modern foreign technology. But the demand for new technology is often unaccompanied by the pre-conditions required by new technology, and this poses a difficult problem. Moreover, because the creation of these pre-conditions often makes it necessary to bring about changes in the traditional values and social organization, conflicts are bound to result. The Japanese experience shows that what minimizes conflict is the attainment of a national consensus, which is a matter of political leadership and of cultural legitimacy.

Confronted with powerful modern technology from abroad, Japan was at first seized with awe and confusion. This soon yielded, however, to a national realization of the need to absorb modern technology. After the initial blind rejection in some corners of society, the country recognized the importance of modern technology. While this both contributed to and resulted from political stability, by the time of the Meiji Restoration, Japan had achieved a high degree of social integration, which was an obvious aid to the formation of a national consensus. The high degree of social integration was due in part to the country's small size and to its several hundred years of complete isolation from the outside world.

In the years before the country opened its doors, the samurai class was the first to recognize the power of modern weapons. Many of the leaders of the new Meiji regime, in visits to Europe and the United States, marvelled at the West's industrial technologies, especially those in marine transportation, railways, and mining.[12] Also, ordinary Japanese, terrified by the contagious diseases that raged as an immediate result of the opening of the country, experienced the miraculous effects of modern medical science and technology and marvelled at this just as much as they did at the strange, new products, devices, and machines from the West.

The Japanese experience began with a naïve, overwhelming encounter with modern technologies, which, however, quickly inspired a national zeal to master and possess them.

Today's developing countries have not had so dramatic and innocent an encounter with modern technologies. They have been using technologies far more refined and sophisticated than those that so shocked the Japanese more than a century ago, but what they have been using and enjoying have not always been suitable for the purposes of their development. Developing countries often lack a national consensus as to the purpose and priorities of development, and their expectations are frequently too great. Finally, the way in which technology has been used has possibly not always been wise.

Definition of Technology

To assure an effective dialogue, it may be necessary for us to agree on the concept of technology. In general terms, it comprises all scientific knowledge deliberately and purposefully used for production, distribution, consumption, and utilization of goods, services, and information, especially that which concerns mechanical apparatus and systems.

This definition of technology, mindful especially of modern industrial technologies, could be useful in considering nearly all the chief problems of development in the developing countries. Of course it may require adjustments, but, in any case, what is important is to enrich our dialogue, more than seeking a definition of technology as an end in itself.

The scientific rationality present in traditional technologies have often been overlooked. Overconcern for the high efficiency of modern technologies may conceal the presence of scientific principles in use by the scientifically illiterate. The Japanese have been guilty of this neglect. Furthermore, an exaggerated reliance on rationalism is apt to oversimplify matters, to elevate technology to the point of being a panacea. This kind of thinking, often ridiculed as being an engineer's way of thinking, lacks concern for important cultural values.

Technology should be evaluated not in relation to its principles alone but in the light of its contributions to the development of the society and country in which it is at work. There are technologies that have served otherwise, of course, but for our purposes, the most desirable use of technology is to serve the national development of the less-developed countries.

The practical experiences of development have shown that because technology—though applied according to scientific principles—is affected by such factors as natural conditions, natural resources and their processing and transportation, and how the technology is used, the attainment of one and the same technological aim may take widely different forms in different countries and regions.

Therefore, unlike science, which is universal in nature, technology consists of intermediated knowledge and skills, which are largely conditioned by geographical, social, cultural, and historical factors. In other words, a scientific principle becomes a technology only when it is intermediated by these factors.[13] Technology becomes stable only after such modification, and is free to disseminate only after it has been stabilized.

The Five Components of Technology

Technology consists of the following five elements, or what we may call the five Ms:
1. Raw materials and resources (including energy): M_1
2. Machines and equipment: M_2
3. Manpower (engineers and skilled workers): M_3
4. Management (technology management and management technology): M_4
5. Markets for technology and its products: M_5

Modern technology must have all of these elements to function properly.

Money and information are also indispensable components. The monetary aspect has been discussed more than adequately elsewhere, so we will ignore it here. As for information, technology requires various types and levels of information, which control and integrate the five Ms. The collection of information is itself a technology that requires a certain level and range of knowledge, a processing system, and a capability to make full use of the relevant instruments and equipment. The more advanced a technology, the greater the amount of information required, and also the higher the level of intellectual capability needed to collect and analyse it.

The effective functioning of technology requires information processing technology, which in turn lends information itself an objective value and cost. Hence, information becomes a central aspect of the managerial policy of an enterprise and, as such, an object of legal protection.

The notion of the five Ms helps us locate problems wherever they might exist in the relationship between development and technology. For example, the same machine will give qualitatively different results depending on the country or company where it is being used.[14] Knowledge of the five Ms is useful in making possible the attainment of the same results from the same machine, or, if not, in finding out what makes it impossible to get the same results.

The five Ms are not present, however, in the same way at all times and places. They exist in different proportions in different countries, enterprises, and factories. This explains the differences among countries in national technology formation. This fact may also be useful for studying comparative national advantages from the viewpoint of an ideal international division of labour in technological development.

From the viewpoint of development strategy, a strategically selected area of technology means a strategically selected industrial sector, which, moreover, takes into consideration the particular characteristics of the goods to be produced. The choice of technology would then need to be made taking into consideration the development levels of allied areas of technology and the links among them. If a technology was chosen where such links were non-existent, they would need to be created, and, in such a case, it would be necessary to conduct a careful feasibility study on the basis of the five Ms as to the appropriate level and scale of the chosen technology. The success of this would depend on the R. & D. capability of the country or enterprise importing the technology.

In regard to the technological self-reliance of a country, native engineers should ultimately play the most important role in R. & D. Foreign engineers and technologists can and should play only a supplementary role. This is an essential finding of our project on the Japanese experience. This is because, in spite of the diachronic, trans-cultural nature of technology, it cannot function independently of the society and culture in which it is expected to function. Only members of that society can make the best use of a technology. In other words, only native engineers can adapt a foreign technology to their country's climate and history, can intermediate, stabilize, disseminate, and, finally, root it firmly in their country.

The Japanese Engineer

Just as technologies have both synchronic and diachronic aspects, engineers may also be so characterized. In some countries, it is necessary for technologists to go abroad for their education, and it is natural that, as a result, foreign technology will be applied in these countries. However, if technologies are to be developed and modified so as to serve national development instead of being always something foreign, the role of native engineers is decisive. Bearing this in mind, let us now discuss the Japanese engineer.

In the advanced and now in the developing countries, there has been a marked tendency for engineers to categorize themselves by function, a trend perhaps paralleling the development of modern technology. In the case of the Japanese engineer, however, he was responsible for a broad range of tasks and functions. Thus, design engineers were also shop-floor manufacturing engineers, and likewise, the shop-floor engineer was expected to develop expertise and experience in the area of design.

Whether the Japanese experience should be made a model in this regard must be left to the developing countries to decide. However, if a technology is being transferred from Japan, for example, it will be necessary to investigate whatever particularly Japanese characteristics might underlie the technology; otherwise, the recipient might find it difficult to see why the technology has failed to perform as well as it did in Japan should such a problem occur.

There is the complaint in many developing countries that Japan and other technology-exporting countries have merely sold their machines and equipment and kept the most important know-how to themselves. These complaints are, in fact, not without foundation, depressing as this is to us and to other investigators. But it must be borne in mind that when a technology is transferred, the culture of that technology is not transferred with it.

One important characteristic of the Japanese engineer is that the functional division of engineering into design, operation, and manufacturing is only relative and temporary. Engineers may be classified by function in a relative sense, but they are not confined to it throughout their careers.

Thus, electrical engineers, mechanical engineers, and civil engineers are not functionally restricted to these disciplines; they are expected to have as wide a range of engineering knowledge as possible, encompassing technologies that reach beyond their own specialties. This has allowed for overlap among the different branches of technology during the period of technological innovation that has been occurring since the 1970s.

Furthermore, Japanese engineers are expected, above all, to be shop-floor leaders, to be able to solve actual problems side-by-side with the workers. They are expected to cover for any shortage of skilled labour and to conduct on-the-job training to increase workers' skills. These are the most important elements setting the Japanese engineer apart.

Because Japanese engineers are usually assigned to design or manufacturing only after on-the-job experience, they are capable of making minor improvements or modifications in production processes and in the design and manufacture of machines to enhance their efficiency and safety. The Japanese engineer first moves horizontally from one sector to another, sectors not necessarily of primary concern to him, on the basis of which he gradually builds himself into a full-fledged engineer. Instead of climbing vertically, then, to become a specialist, the Japanese engineer becomes an all-rounder in technology.[15]

Therefore, when we talk about the characteristics of Japanese technology, we are also referring to the peculiar way of training engineers in addition to technology control and management.

The point here is not that this approach is better than others; the important thing is that Japan became self-reliant in technology aided by this particular way of training its engineers. That the birth of this type of engineer took place in the initial stages of industrialization, when the absolute number of engineers was small, may prove to be useful information when studying the problems of development and technology.

It might be of interest to add that, even in the days when there were few engineers, they did not occupy very high positions at their work places, and their social status was not high. Their salaries were relatively good, but they usually had only limited power. Perhaps this is a phenomenon peculiar to a technologically less-developed country; we met many engineers in the developing countries we investigated who had much to complain about because of their social positions. Their problems were similar to the ones Japanese engineers once faced.

Japanese trained engineers first began assuming leading corporate positions in the 1910s and 1920s, a period when the country had succeeded, for the first time, in developing indigenous modern technology. Engineers began to widely occupy the highest executive posts in corporations only after World War II, when corporations became owners of technology and possessed what amounted to armies of engineers.

In considering the relationship between scientists and engineers in the initial years of industrialization, attention should be paid to the following two relational aspects.

The Relationship between Engineers and Techno-Scientists

The two now belong to categories relatively independent of each other, though they were interchangeable at earlier stages. The techno-scientist's primary job has been to collect technological information from different parts of the world and analyse it. Although this group has formed a key part of the core for technological development in Japan, their work has centred on basic research and experimentation, being removed from shop-floor operations, though the distance has been shortened somewhat in recent years.

Techno-scientists have long served as advisers to the state in the formulating of Japanese science and technology policies, whereas Japanese engineers began only recently to make themselves heard from the shop-floor in terms of national technology policy.

Finally, at the earlier stages of industrialization, techno-scientists made great efforts to train successors as well as skilled workers in new fields. Their contributions to the education of engineers and skilled manpower since the end of the last century have created a bridge between science and technology; bonds like those between a master and his disciples were forged between techno-scientists and engineers (though new problems arose later). Such bonds, and the camaraderie among fellow-students, aided the growth of industry-university co-operation, especially in the areas of technology that played a leading role in the formation of a national technology network and in which Japanese technology has risen to world prominence.

Workers' Attitudes towards Engineers

In the earlier stages of industrialization, when the number of engineers was small, a higher or specialized education was evidence of one's family's elevated social status. Not surprisingly, engineers enjoyed more favourable circumstances in terms of both the status and conditions of employment; yet in spite of that—or because of it—engineers were always the shop-floor leaders, and the workers, though sometimes the envious subordinates, held those superiors who could competently address and solve their problems on the shop-floor in high regard. If their superiors were incompetent, the workers would remain obedient, but, at the same time, lose the incentive to work hard, for the quality of shop-floor leadership had a direct effect on worker safety, productivity, and wages.

Japanese workers dislike having designs and specifications changed while work is under way and will sometimes even openly oppose any such changes. The European-style functionalistic attitude of workers, who may not object to such alterations so long as they receive their due pay, is rare in Japan, where the attitude of both workers and engineers towards technological skill and competency is strict. Even labour union leaders are not likely to be elected merely on the basis of their capabilities to organize and bargain. In the early years of the labour movement, all union leaders were workers of outstanding skill. This reveals the great importance Japanese workers place on labour and skill, not unlike the value an artisan places on craftsmanship.

It was natural that such ethics should have been reflected in both the consciousness and roles of engineers, and thus the Japanese engineer was formed.

The Five Stages towards Technological Self-Reliance

The Japanese experience has shown that, if technology transfers are eventually to lead to technological self-reliance in a given nation, it must create its own style of integrating the five Ms and its own corps of native engineers.

The Japanese road to technological self-reliance was marked by a series of painful efforts and several stages extending for more than a century.

Although the work of our project was based on case-studies in diverse sectors of industrial technology, we may generalize them and divide the history of development of modern technology into the following five stages:

1. Acquisition of operational techniques (operations)
2. Maintenance of new machines and equipment (maintenance)
3. Repairs and minor modifications of foreign technologies and equipment, both in the system and in operations (repairs and modifications)
4. Designing and planning (original design and creation of a system)
5. Domestic manufacturing (self-reliance in technology)

To attain complete self-reliance in technology, none of these five stages may be skipped. The advantage of a late starter is the possibility to economize on the time, money, and energy to be spent at each stage of development. It is not necessary, nor is it possible, for every country to attain complete self-reliance in technology or to develop all areas of technology in an autarkic manner. What a late starter must do is choose a sector of technology in which it has an advantage, taking into account its own development needs and the optimal types, levels, and scales of technology.

We have outlined these five stages of technology development because, as many shop-floor engineers and historians of technology have pointed out, there is no such thing as a leap in technology. We hope this breakdown will serve as a useful tool when considering the relationship between a nation's development and technology. Judging from discussions of the subject, it seems there is a tendency for the argument to jump from stage 1 to stage 5, to the problem of manpower or to the politics of technology, with little attention given to stages 2 and 3.

To clarify what is presently the most urgent of the technology problems of each developing country, it might be useful to combine the five stages with the five Ms of technology. Modern technologies are interrelated. Therefore, even if national self-reliance in each transferred technology were attained through the five stages, the path would not be straight, but would, rather, follow a spiral course to self-reliance. Thus, a technology enclave, a technology transfer at the hands of a transnational corporation, having no intention of developing the related technology outside its own sphere, would not contribute to the formation of a national technology network in the host country, and we can, therefore, disregard cases of this kind here.

Modification of the Five Stages

The five stages of technology development may need some modification for selected technology transfers. For example, reversing the order of stages 2 and 3 may be appropriate: this may apply to transfers in a country that has reached a level of technological development characterized by comparatively simple machines and facilities. One of our collaborators, Professor Hoshino Yoshiro, drew our attention to the existence of this situation in China.

According to Hoshino, where individual skilled workers are capable of improving machines, for example, there may be a lag in establishing a national maintenance standard, and this would bear directly upon stage 4. Therefore, the order of stages 2 and 3 might be reversed.[16]

On the other hand, in such a complex technological system as a manufacturing plant, the maintenance technology should be established first. Repairs and minor improvements or modifications will be possible only once that has been done. In countries unable to manufacture basic automobile parts and accessories, for example, maintenance technology is vital.

To take examples from Japan, a world-famous clock and watch maker started business as an importer and repairer of foreign clocks. A well-known manufacturer of electrical appliances began as the engine-repairing section of a mining company. To begin with maintenance and repairs and go upstream to higher technological areas is a quite ordinary way to accumulate technological capacity.

In our field-work at several factories, we found that maintenance training through the periodic dismantling of machines and equipment for overhaul is very important to maintain the efficiency of machines. We also learned that the dismantling itself was regarded as an apt index for evaluating mechanics' skills. In recent years, however, several chemical plants have increasingly entrusted maintenance work to outside specialists; some engineers have criticized this division of technological labour, saying it will lead to a diminishment in workers' skills. It should be noted, though, that there are now cases in some areas of technology, especially in those of already mature technologies, where skills in maintenance and repair do not always lead to the enhancement of operational skills. The question of whether to build up a maintenance technology for each industrial technology or to leave it to outside specialists should be a matter of choice for each country.

There is another important consideration regarding the five stages of technology development. The five stages as a whole have information as a common factor on the one hand and manufacturing capability as a common factor on the other, just as the five Ms of technology have financial resources in common on the one side and information on the other. That is, the same technology may be observed in use at different levels of the five stages in different countries because of their divergent information and manufacturing systems and capacities. Therefore, it might be better in some cases to order machines and equipment from foreign manufacturers if they can design them to suit the user's domestic conditions, instead of trying to manufacture them domestically.

But constant reliance on foreign technology is undesirable. Developing a national/local capability in maintenance and repairs and in making modifications is important because foreign technologies often lack uniformity in their standardization; this is a result of the society in which they originated. Nevertheless, it cannot be denied that establishing a stable connection with specialized foreign enterprises may be a choice a developing country wishes to make.

Three Elements of Technology Management:
Eliminating *Muri*, *Muda*, and *Mura*

Modern machines and tools have become increasingly maintenance-free. This has been made possible through the development of stronger materials. But unlike their predecessors, modern equipment is less flexible with respect to function and durability. In any case, handling skill, support services, energy supply, etc., still have an effect on the output and durability of the equipment.[17]

Even when the technology is embodied in the final goods, in an automobile or a machine tool, for example, it—and the skills that applied it—will affect the output and service life of the goods. In production technologies, handling and control affect the efficiency and quality of the products even more, and, thus, the quality of management in the enterprise that owns the technology is of great importance.

For example, at the Amagasaki mill of Kobe Steel, there is the slogan "Eliminate '*Muri*,' '*Muda*,' and '*Mura*'" (*muri*—to overwork or do something forcibly; *muda*—to waste, diseconomy; and *mura*—irregularity or inconsistency). This watchword, emphasizing rationality, safety, efficiency, economy, and a high standard of quality control, pin-points the essence of factory management and technology control today.

During our visit to the mill, I noticed that, although most Japanese were impressed by the slogan, representatives from developing countries seemed unmoved. The difference in response seemed not to indicate any personal disagreements; it merely reflected the state of technological affairs in the visitors' home countries. The questions they asked the factory staff centred mostly on such matters as the strategic placement of the steel mill's technology in regard to defence from sea-based attacks, labour management, and QC circles—matters taken for granted by the Japanese.

In addition to concern for cost, quality, and security, prompt delivery might also be mentioned, for it is what clients especially demand of producers of intermediate goods. In automatic production, for example, where assembly lines handle hundreds of component parts and accessories, delivery dates and times have a decisive importance in maintaining productivity. The higher the level of technology, the more important the punctuality of delivery, because such technologies are dependent on complex, wide-ranging support sectors. A refined delivery schedule enhances performance, and a poor schedule creates operational diseconomy. Toyota's "just in time" delivery system (the Kanban method), for example, eliminates the necessity to stock many spare parts and to maintain large warehouses, which, in turn, reduces production costs considerably.

Obviously, this system requires punctual deliveries to maintain a high level of productivity. Since punctual deliveries over long distances depend on good communications, information, transportation, and other services, some of which are beyond the manufacturer's direct control, there must be some safeguards against unexpected snags and losses. An essential factor in technology and factory management is to be constantly prepared to overcome any crisis that threatens to disrupt constant full-scale operations.

Factors such as these constitute the heart of technology management and control, though they are often overlooked by technology users, and this is one difficulty Japanese technology exporters have sometimes encountered, even in the industrialized countries.

In the light of the experiences of the developing countries, the five stages of technology development might be supplemented by another one preceding the first; namely, a careful assessment of the costs and benefits likely to result from a transfer of technology. Even should a country find it necessary to transfer a technology, it might lack the right conditions to do so. For instance, where large-scale equipment must be introduced, the necessary port and transportation facilities may be lacking. Feasibility studies carried out before technology transfer, however, might reveal important difficulties. For example, even if it is found indispensable to prepare or build infrastructural facilities to make a transfer possible, it might be determined that, if the infrastructure were built, it would be needed only at the time of the transfer and would later prove useless. The costs, including the cost of maintenance and administration of the infrastructure, would then be ruled too high to carry out the transfer.

Many developing countries are also often obliged to import technologies that are too large for their needs, but they would have to have an exceptionally high level of R. & D. capacity and engineering ability—both usually rare in developing countries—to modify the scale of the technology. As a result of this difficulty, technology transfers to developing countries often prove unsuccessful.

The question of economic rationality in the choice of technology also arises. If the technology importer is a private enterprise and alert to economic rationality, it will likely modify its transfer policy to avoid diseconomy. However, the transfer of a technology to answer the technological needs of the state may present a different case.

The needs of the state have often been those of the political or administrative élite, who tend to pursue only the latest technologies and equipment. As a result, the maintenance, management, and the products themselves are apt to be inferior, yet expensive. This can be expected because the élite lack the required expertise and an alertness to the aforementioned three elements of technology control.

Japan is a good example of this. Many of the early, state-run factories using newly introduced foreign technologies proved unsuccessful. Once in private hands, however, they exhibited significant improvement in both managerial and technological capabilities.

Kamaishi Ironworks, for example, modified the foreign technology it had taken over from the state to suit the raw materials that were locally available; further, it reduced its scale to stabilize operations. Two important reasons for the failure of its predecessor were the original design by foreign experts, who had introduced their technology to Japan without modification, and the Japanese government's policies on technology. It was Japanese engineers who had to remove the difficulties.

We heard similar stories in the developing countries we visited. The quality-control movement that Japanese experts tried to introduce there did not at first prove successful, but after a change in management, it was learned, the movement was proposed anew by native engineers and successfully realized.

A Chinese scholar stated that he was impressed that in Japan a thorough preliminary cost-benefit assessment was usually made before a technology transfer. This is taken for granted in Japan, but the remarks remind us that technology transfers often occur under the auspices of development aid, in which cases primary importance is attached to government needs, and the analytical assessment of the technological and economic rationality of the transfers is ignored. The urgency of development tends to justify these sorts of technology transfers to developing countries that are unconcerned with techno-economic rationalism. It is important to remember, however, that modern technology is primarily based on economics, even if it cannot be free from national and international politics.

It may also be added that technology, like economics, will be sure to stagnate without free and fair competition, and so will the quality of engineers. Where engineers and skilled workers are few, the ability to develop technology is difficult to foster. In Japan, the technologies that remained long under government control developed and spread more slowly than those that did not.

For example, the spread of telephones under government administration in the 50 years they were in use before World War II was incomparably slower than in the 30 post-war years when the telephone business was in the private sector. In pre-war days, because extension of telephone lines was very slow, even having a telephone was considered a symbol of wealth and social status, and telephone owners tended to support the restricted availability of telephone access. The extension of telephone lines was made both possible and necessary through innovations in telephone technology and the increased popular demand for telephones, resulting from the greater income of the people and changes in their life-styles. In sum, state-run businesses tend to be slow to respond to a nation's needs.

Though bureaucratic control of technology may cause greater difficulties than market-oriented, privately run, business-oriented management, the bureaucracy itself is not the same in different countries and in different times. The intention here is not to suggest that bureaucratic control is always inefficient and uneconomical; nevertheless, a monopoly on technology, whether state or private, is not good for its diffusion.

Part 2

Case-Studies

4

The Importance of Case-Studies

There are many possible approaches to discussions of technology, and yet it must not be construed as an abstract subject. Technology is a concrete ingredient of daily life and must be dealt with in a concrete way. Therefore, the most meaningful dialogue on technology is initiated and developed in reference to actual cases.

In our own discussions we have encountered such ideas as "technology civilization" and the "nature of technology." Our present study, however, has no direct bearing on these and similar notions. Our main theme is development and technology. The urgency of this subject is found in the torrent of many small but real problems that have gone unsolved and not in philosophical discussions of the nature of technology; we have been overwhelmed by the urgencies of real, everyday problems that must be successfully solved.

Rather than permanent solutions, we are interested in provisional solutions of smaller problems that exist now. Admittedly, however, even minor solutions cannot be expected immediately. Indeed, minor solutions, limited as they are, will surely generate new problems. We know that we may need to be satisfied with minor solutions to problems that can be anticipated when a question is raised, and we may even need to regard such solutions as final.

The methodological dialogue proposed by the United Nations University represents a challenge to the current situation in which problems of development have been confused and their solutions only groped for. The intention of a methodological dialogue is good, but the results have been modest, and a full appraisal remains to be reached. Furthermore, no grammar seems yet to have been established for methodological dialogue premised on the individuality and equality of the participants. We have already heard voices of disappointment, the reason for which is understandable. Certainly, methodological chaos has thus far dominated the dialogue. However, the disappointment characterizes the current situation of the development problem because, paradoxically, it expresses hope and expectation. There is no quick or

65

universal remedy that can be applied either to development or to methodological dialogue. Nevertheless, dialogue is meaningful because it promises an opportunity for all concerned to better understand the problems. This is the positive significance of dialogue.

We are in desperate need of detailed information on the problems of technology in developing countries in order to see the problems in their real context. Although there is no reason why the way in which we define a problem should be the same as or unified with all others, the factual information of case-studies must be provided for meaningful dialogue. Cases that cover a wide diversity of levels, ranging from factories to space technology, have been brought into our dialogue; however, we will confine ourselves first to problems at the level of the nation-state, inasmuch as the subject of development is the nation-state and development is a matter of its sovereignty. We then move to the level of the factory.

Although Japan early constituted a nation-state, conscious awareness of this on the part of the Japanese people came late. Until Japan was forced to open its doors in 1854, the average Japanese regarded his or her microcosmos as the universe. Although Japan had attained a high degree of social integration, unification as a national society came about only through industrialization, a process that began with the forced opening of the country. The 200 or more years that preceded the opening and the provincial mentality of this period, as well as the power structure established on the basis of this mentality, will be discussed in this section in the many diverse ways they apply to technology in the 1980s.

Japan moved from its feudal system, made complete with the closing of the country, to a state following the historical model formulated by European historians. In nominal terms, however, power and authority existed collaterally. The new government had achieved the unity of a nation-state, but in its tenth year (1877), the country was hit with a civil war. This furthered the economic confusion that had existed since the opening and confronted the newly born state with a threat to its existence.

Colonization was another serious threat. Foreign armed forces were permanently stationed at Yokohama; facilities were provided at the expense of the Japanese government, and extraterritoriality was granted. The new government even mortgaged the Yokohama Customs House to repay a loan it had inherited from the old shogunate government. As a result, there was no customs autonomy.

To maintain political independence, the new government had to repay its foreign debts. This made economic development essential, and the only available option was the introduction of European technology to incite an industrial revolution. Herein lies the relevance of the Japanese experience to development.

In Japanese academia, modern Japan has not been examined in terms of third-world development; it has been directly compared with only other advanced countries and not with those in Asia, Africa, and Latin America.

The conditions for international co-operation were not as well established when Japan began its development as they are today, and Japan had no neighbouring countries with which to co-operate. China's defeat in the Opium War frightened Japan's intellectual class; China had always been respected in Japan as a political and economic model, and this turn of events thus forced a reassessment.

Japan attempted negotiations with the Li dynasty of Korea so that it might cope in alliance with the new international environment, but Korea refused to open its doors. In an unexpected response, it even asked Japan to regard Korea as a superior nation, and the negotiations failed. Japan was thus isolated in the Far East and was put in the position of having to modernize on its own. Consequently, attempts at development were impetuous and met with a series of failures.

The technology transfers to Japan fell short of their goals because of shortsightedness and naïvety. The lesson sorely learned was that future transfers had to be made selectively. So after several overly ambitious and optimistic experiments, technology transfer was re-initiated on a scale that was more realistic and rational.

Where technology was independent of control by the government and politicians, the five Ms were prepared and the question of technology settled. The five Ms and the establishment of technology were realized first in such light industries as textiles and food processing, later in the mining industry. Heavy industry existed only on a small scale, and it was not the leading sector.

Motivated by the needs of national defence, the Japanese government encouraged the manufacture of iron for shipbuilding and for producing arms and ammunition; the government also promoted the development of iron mines for iron manufacturing and the construction of railways and ports for iron-mine development. The process represented the reverse of that of modern Europe, though in overall industrialization, Japan trod the European path as it moved from light manufacturing to heavy industry.

But Japan accelerated and shortened the process. One factor in the acceleration was the government's policy of encouraging industry. This policy was co-ordinated by the national ministries regarding implementation, but, more significantly, it was implemented both for the central government and for the prefectural and village governments to assist on all levels in survival and development amid the serious economic upheaval they were experiencing. The name of a clothing manufacturer, Gunze, literally "district guideline," provides perhaps an amusing display of the strategy.

The policy of encouraging industry was not first formulated by the Meiji government; such a policy had been implemented by the feudal clans under the Tokugawa shogunate (for example, one by Yokoi Shonan [1809–1869] of the Fukui clan). However, the Meiji government developed the policy on a nation-wide scale.

In developing a military industry to meet state defence needs, the overwhelming technological gap could not be narrowed by only the purchase of equipment and machinery. It was necessary to learn how to operate the equipment, and to manage its maintenance and repair.

But, before the needed equipment and machinery could be introduced, foreign currency was required, and for this, the technology for silk reeling was brought into government-operated plants to provide an export product for foreign markets. But in this we see a paradox: the promotion of agriculture was necessary first to make industrial technology possible.

In the mining industry, copper-mining developed favourably by using transferred technology until the end of the nineteenth century. As in coalmining, the copper was not for domestic consumption but for export.

It is inevitable that late comers in industrialization will aim at increased production or development of agricultural products and raw materials for export. Only at a later stage can the agriculture of a country supply raw materials to its own manufacturing industry. Following this model, it is important that each agricultural division be looked at in terms of its stage of development and that an assessment be undertaken of what industrial technologies have been transferred and established in relation to other industrial divisions, in order to take action toward answering development needs.

5

Urban Society and Technology

The City and Technology

It is important to include the problems of cities when examining the problems of technology transfer and development. It has been the academic practice to treat these three areas separately—though they are interrelated in complex ways—following the tendency to reduce problems to the level of individual disciplines. To remedy this, we have formulated the themes of "urban society and technology" and "rural society and technology" in examining individual sectors of industrial technology. Concentrating on these problems is perhaps more relevant to development than whether we have obtained sufficient results to solve them directly.

The urban problem is today a global problem, encompassing both the North and the South, but the problem differs in content and structure between the two areas. The South faces more complex and difficult problems in employment, transportation, housing, health, social security, and similar issues. The measures needed to solve these problems cannot be the same for each country. In the context of development—exemplified by such problems as the world's highest rents and land prices—Japan has solved these problems only partially; nevertheless, a few illustrations from the Japanese case may be of interest in our dialogue.

The relation between the city and technology in the context of development is our concern for the following, related reasons:

1. Technology and its transfer are essential for development.
2. Modern technologies, hard and soft, are mutually interrelated, each tending to concentrate in places where related technologies and supporting services are available.
3. The city is where technology, service, and information are centred. In other words, where technology, service, and information have been centred and accumulated is or will become a city.

69

Modern industry has developed and given rise to new cities. However, industrialization and modernization are defined, their processes are sure to generate urbanization, which is a prerequisite for industrialization and modernization that will, in turn, increase urbanization—as the Japanese experience demonstrates.

For example, such public services as transportation and electricity, essential for the modern city, will become available where a certain level of urban population has been reached. Developing countries have arrived at the stage of urbanization that constitutes a prerequisite for modernization and industrialization. Therefore, if priorities are set and wise technological choices and timely transfers of additional technology are continually made, these countries may anticipate success in their work to solve the problems they are likely to face.

Awareness of the particular phase of urbanization in which a country finds itself may be useful in making plans for that country. Urbanization has three phases:

1. Population expansion in existing cities.
2. Increase in the number of medium- and small-scale cities.
3. Development of the division of functions among cities and the nation-wide formation of hierarchies to parallel the levels of these functions.

These phases correspond to the process of development, starting when the technology and the organizations for technology are first scattered, moving to when they are brought into some regional concentration, and finally to when they are integrated into a single entity.

In terms of this framework, many developing countries have completed the first phase, but not sufficiently the second and third phases. This means that the national network of technology in each does not yet cover the entire country and that the level of social integration is not high and the social structure is not as solid as it should be. Further, because of the rich diversity of local culture, the network for administration (national bureaucratization) does not function efficiently. This can, of course, prove both advantageous and disadvantageous to development.

In this context, by the middle of the nineteenth century, Japan had a network of more than 200 cities with populations ranging from 10,000 to 1 million. The functional hierarchy among cities had been completed by this time, and they were connected by all-weather roads and water-borne traffic. The Meiji Restoration included the reorganization of more than 240 small administrative units into 50 larger ones, and the new central government could recruit bureaucrats for administration not only from the ex-samurai class but also from other classes. This guaranteed it a cadre of leadership.

The Primate City

One characteristic of the process of development in today's non-Western world is that, while capitals have expanded to extraordinary sizes, the development of the secondary and tertiary cities has lagged far behind.

Tokyo is Japan's primate city. Its population at the time of our study was 11 million, only about 10 per cent of the total population of Japan. Within 30 kilometres of Tokyo is Yokohama City, which has a population of 3 million. The twenty-seven towns and cities in the Tokyo metropolitan district and the several neighbouring cities, such as Kawasaki City (more than 1 million population), boast extensive public transportation networks, bringing millions of commuters into the city each day. For this reason, the daytime population in central Tokyo exceeds 14 million people. The metropolitan area of the three large cities of Osaka, Kyoto, and Kobe, in western Japan, is less than 70 per cent as large as metropolitan Tokyo. But in the centres of both metropolitan areas, the night-time population decreases drastically, making them gigantic hollows. This differs markedly from what one finds in cities of the third world.

As the primate city, Tokyo constitutes a gigantic metropolitan area. It differs from large third-world urban centres because it sits atop a hierarchy of as many as 10 cities, each of which has more than 1 million population. Next on the list of cities in Japan are 10 with populations of 500,000 to 1 million and 38 with 300,000 to 500,000. It would be advantageous for the development policies of developing countries to promote the development of small and medium-sized cities.

However, while technology investment is influenced by related industries or supporting services, investment in the development of local cities faces its own set of limitations and urban investment tends toward the primate city. The renewal and redevelopment of facilities in the primate city, which, because of their scarcity, may have worn out through excessive use, is in competition with the development of local cities.

The formation of a primate city or excessive urbanization generated by population explosion and a low degree of industrialization have brought confusion to the cities. The phenomenon symbolizes the disorder of the entire national society at the initial stage of development. And although such confusion can lead to the dissolution of the national society, it should not be assumed that it does so automatically.

In the Japanese experience, excessive urbanization in pre-modern Japan occurred together with a population explosion. In the mid-nineteenth century, the political centre of Japan, Edo (present-day Tokyo), had a population of more than 1.1 million; the second largest community, Osaka, had 400,000; the third largest, Kyoto, had 300,000. Edo was indeed the primate city of Japan, and represented the country's political centre, with Kyoto the centre of sovereign authority and Osaka that of the economy, the three thus sharing the major functions.

The Meiji Restoration meant the unification of political power and sovereign authority in the new capital, Tokyo. The industrialization policy of the Meiji government brought about a new economic centre in the capital to compete with Osaka, but Tokyo could not easily dominate Osaka, and for a long time the two cities shared national economic hegemony. As industrialization progressed, however, Tokyo gradually established a predominance over Osaka.

The Inhabitants of Tokyo

The change from Edo to Tokyo signified a change in the political power and the polity. This change first brought about a decrease in population. Following the transition, the 1.1 million population peak of the Edo period fell to approximately 500,000. Some 20 years later, however, when the new government had become stabilized in the 1890s, the population again reached 1 million. Only 10 years later, it approached 2 million, more than 1 million of which represented an influx from outside.

The reason for this great influx was change in the villages. The land and taxation systems were reformed, freedom of occupational choice was guaranteed, and freedom of movement and residence was authorized. People were liberated almost overnight from the former feudalistic restraints. Many of the people moving in, however, were poor peasants who had been uprooted. Their poverty was a result of several things: major economic change, especially the inflation that had occurred following the opening of the country and the civil war; natural disasters that were induced or worsened by the administration's mismanagement in forestry conservation and riparian improvement; and the spread of contagious diseases.

Records kept by Europeans who visited Tokyo immediately after the Meiji Restoration indicate that people were extremely poor; many were half-naked and living in shanties. While most of the houses of former samurai were left empty, many people lived in shanties along the streets, the only place they could engage in peddling, to which they were restricted and for which no alternative job opportunities were provided. The shanty towns usually grew up within a range of 2 to 5 kilometres from the entertainment or business centres, areas located within the metropolis much the same as they are evolving today with populations of new inhabitants in cities of the third world.

In nineteenth-century Tokyo, the water-supply and drainage systems were not well established, and the water-supply network, which had been constructed in the Tokugawa period, fell to ruin when maintenance on it was discontinued. Underground water was abundant, however, and people could get water from shallow wells (Kosuge 1980). On the other hand, poor drainage facilities caused Tokyo to suffer repeatedly from water-borne diseases. The spread of cholera after ports were opened to foreign trade killed more than 100,000 people on each of several outbreaks that struck every several years until the beginning of this century, a period of more than 40 years. Under the unequal treaties forced on Japan by the Western powers, the government could not take preventive measures against epidemics because the diplomatic representatives of these powers opposed these measures. The most frequent victims were the lower-class city dwellers.

As the eighth governor of Tokyo (in office from 1883 to 1886), officially declared, "the roads, bridges, and rivers come first; the water-supply system, housing, and sewerage are secondary." The Meiji government gave priority to the modernization of industry and the military. Public welfare in the cities was of only secondary concern to the government, and Tokyo thus added new urban problems induced by industrialization to the traditional ones it had inherited from the Edo period (Ishizuka 1979).

The Meiji élite believed that Westernizing the capital was essential in contributing to a revision of the unequal treaties with the Western powers. The new government office centre of stone buildings and the construction of brick buildings in the Ginza area were planned, moreover, to make Tokyo a city safe from fire. (Edo was the victim of major fires about every three years, and insufficient fire-prevention facilities had hindered the construction of full-scale wooden buildings.)

Although the more fire-resistant construction materials offered protection from fire, the change was not favourably accepted by the people because these materials did not fit the natural environment of high temperatures, high humidity, and frequent earthquakes. Consequently, the Ginza brick-town plan was never completed. In a rather meandering fashion, the Ginza of today was formed, and Tokyoites today little realize the fashionable Ginza was once a slum area.

In general, urban planning in Japan until the defeat in World War II consisted of industrial and transportation network development for national defence, leaving the housing problems in the hands of the private sector. Real-estate dealers operating rental houses reigned in the slum areas. Residents rented tenement houses partitioned like the mouth of a harmonica, and the rent was collected daily. They lived literally from hand to mouth. Rain prevented many people from working outdoors, often requiring them to pawn their little bit of furniture and working tools to buy sustenance. Near the slums were prosperous pawnshops and street stalls selling food made by re-cooking the remnants of meals from school dormitories and hospitals.

The Poor

Although most of the population in Tokyo was made up of impoverished newcomers, they were not a uniform group, but comprised three subcategories:

1. Low-Income Artisans

This group was represented by such outdoor construction-related workers as carpenters, plasterers, stonemasons, and gardeners, and by such skilled craftsmen as gold- and silversmiths, furniture makers, and tailors. These people retailed or peddled their own products. Owners of small shops with petty capital were included in this category.[18]

2. Paupers

This group included rickshaw men, representative of the new urban labour that had come about since the 1880s, daily labourers, and unskilled outdoor manual labourers. (The rickshaw men consisted of such types as self-employed, retained, renter-live-in, etc. The size and stability of income varied according to the order in which they are listed here.) The rickshaw man's income was sometimes above the average income level of low-level urban workers; but the work was so hard they could seldom continue their work much beyond middle age.

3. The Destitute

This class consisted of persons engaged in light labour, such as street merchants, peddlers, street performers, beggars, rag-pickers, and vagabonds. Some characteristics of this class were that there were no age and sex distinctions by type of job, the jobs were ill-defined, and the workers often had no permanent residences. In other words, this group constituted the lowest stratum of urban society. The disabled, the old and infirm, and the physically handicapped who could not manage even light labour were included in this group.

Groups 2 and 3, and sometimes a portion of group 1, could be referred to as the odd-job stratum of the cities; but its constituents are much more varied than the so-called lumpen proletariat. This class may be considered similar to the populace often referred to today as the urban informal sector. In any case, though this group included criminals and slum dwellers, the class as a whole cannot appropriately be characterized as antisocial. Here we find a fundamental difference from the exclusive, closed, small social group classically seen in an industrial society. Applying to this group the terms "slum of hope" or "slum of despair," or both, does not contribute to a true understanding.[19]

Category 3, the destitute, clearly needed protection and relief. And yet, the Tokyo city government underestimated this population category, placing it at about 10 per cent of the whole (Nakagawa 1982). The authorities believed that welfare policies directed at this class would cause a new inflow of the same class into Tokyo, and because the whole society was in transition, this would lead to bankruptcy of the city budget and spoil the will of these people for self-help.

In the 1890s, members of what we are calling the odd-job stratum had an average income of four yen (about four US dollars), and needed 70 per cent of this income for food and 15 per cent for rent. These people, on the edge of starvation, were seen not only in the principal cities but also in medium-sized and small municipalities and in rural villages. Some were employed by a village or jointly by several villages to fill the most menial jobs.

Today, according to one report, nearly 40 per cent of the people in South Asian agricultural villages consists of a non-agricultural population, a class that corresponds to the odd-job stratum described above in reference to Japan. Because of the population explosion and the limited means to sustain a growing population in the villages, impoverished peasants flow into the cities and fill the odd jobs, a phenomenon that could be explained by the widely accepted push-pull theory.

As a corollary to this, Yokoyama Gennosuke stated in his book, *Nihon no kasō shakai* (The lower strata of Japanese society), that, once a population has flowed into the cities, it very rarely returns to the villages whence it came.[20] Although Calcutta reported that in the 1960s some 200,000 people migrated from the city during harvest season and then returned when the season ended, this was a temporal outflow, a manifestation of a so-called floating population, a phenomenon which, assuming Yokoyama is correct, was for the most part non-existent in Japan at the time.

According to Yokoyama, Japanese cities at the end of the nineteenth century had the capacity to absorb the odd-job stratum, an ability comparable to contemporary third-world cities. A possible explanation for this might be that the initial stage of industrialization is accompanied by a population explosion, and the number and size of slums—not necessarily centred on the primate city—are proportionate to the size of the cities. An enlarged slum provides specialized minor jobs, and the demands for diversified goods and services support poor people and stabilize their livelihood at a low level.

Sojourners

At this time there were a great many adult male lodgers and temporary residents (sojourners) in Tokyo. In 1869, immediately after the Meiji Restoration, the adult male population of Tokyo (250,000) almost equaled the adult female population (260,000). Twenty years later, however, the male population had increased by more than 100,000—most of this represented by the sojourner class—though the total population of the city remained the same as in the Edo period. Most of these sojourners were to be found around the tradesmen's houses in the commercial and industrial areas and along the nearby alleys where the low-income artisans and paupers lived, and this was closely related to the odd-job nature of their work: the odd job that the sojourner relied on for his living could not be found elsewhere. Their employment opportunities were sharply limited, and, as a result of these circumstances, this group lacked the means to settle down and form families.

Not all of the sojourner population, however, was to be found in the low-income stratum. Many bureaucrats in the new government, who were high-income earners, had come to Tokyo in its early days, leaving their families in their home towns. Even when the families joined them, they did not give up their permanent domiciles. These bureaucrats provided support to relatives and others who did leave their home towns for Tokyo or, if not, help in securing employment. It was regarded as the social responsibility of those who were successful to provide for their home-town friends and relatives.

The relations thus formed on the basis of a common origin or through mutual reliance helped connect Tokyo and the local areas and sometimes contributed to the creation of strong factions. This was especially notable in the bureaucracy, particularly among high military officers who came from a limited number of clans. With the increase of population in specialized and skilled occupations, however, the principle of personnel selection based on academic background or general merit gradually became established. Nevertheless, all other qualifications being equal, in terms of reliability and expectations, priority in hiring was given to persons from the same province and the same universities, the academic cliques duplicating the cliques based on place of origin. Long-term investments for the development of education and human resources were influenced by the formation of these cliques.

Although support for or opposition to the new government had been based on one's provincial affiliations, the basis for deciding loyalty slowly moved from provincial loyalty to actual merit resulting from the acquisition and use of new technology. And the changing basis for evaluation was expanded from the political and military arena to foreign trade, the arts, and technology in the service of national development.

Formation of the New Middle Class

The great effect industrialization had on urban society was the dissolution of the pauper class. New occupations (such as florists and makers of footwear and bags) were created, and the living standard of skilled artisans was gradually improved and stabilized. The rickshaw men and outdoor labourers of the former pauper class established families, thus creating a new phase in urban society.

When urbanization entered this new phase in the 1900s, the persons who later were transformed into industrial workers were born into the urban lower classes. As subdivision in manufacturing progressed, through a system of subcontracting that resulted from a breakup in the production process, some skilled workers became owners of small factories or of part of the shop-floor production. As larger industrial networks formed, members of the urban lower classes moved to the areas surrounding the big factories, thus transforming towns into strongholds of factory workers and their families. At the initial stage, before this development, the factories themselves were located near the slums, on land that was cheap and where the recruitment of labour was easy (Ishizuka 1980). Cotton-spining factories were the most typical example.

Simultaneously, a new middle class was emerging, including low-level civil servants such as school teachers, public employees, railway workers, and low-ranking professional soldiers. The government provided them with housing and long-service pensions. Although their levels of income were not so different from those of skilled workers or owners of small shops, their social consciousness and life-style set them apart. For people in this class, even when the income of the head of the family was too small to support the family, the wife or other family members hesitated to work outside the home, and the deficit would thus be supplemented by part-time work done at home.

Above this class lay a thin upper-middle class: owners of big shops, academically trained professionals, independent business owners, managers in big enterprises, and high-ranking bureaucrats. An element that differentiated this higher middle class from the new middle class was that, besides houseboys and maids, it had unmarried female domestic labour, called "housekeeping trainees" (with the housewife as the teacher).

This domestic labour force, especially the trainees, gradually disappeared after the establishment and spread of the school system. World War II made it impossible for households in this class to keep domestic servants, male or female.

The expansion of the old middle class and the formation of the new middle class provided people of the low-income artisan stratum (group 1) with a stable market for their products and services. A great many new occupations came about with the appearance of small-business owners. The odd-job stratum did not disappear as an entity, but reformed to produce a new supply of industrial workers. This and the new middle class, which formed and expanded in parallel, signaled a new phase of urban society brought about by industrialization.

Urban Life

Tokyo consists of two poles in terms of society and culture: Yamanote (the hilly, uptown part of the city) and Shitamachi (downtown), geographical references dating from the Edo period. As in the distinction between town and village, it is difficult to precisely distinguish the two. In Tokyo, 10 times bigger than Edo, the two are intermingled and multilayered.

However, some distinction can be made regarding differences in life-style, values, and especially in what might be termed social aesthetics. In general, Yamanote people are progressive, Western-oriented, individualistic, better-educated, professional or managerial types, and speakers of standard Japanese. The disposition of the residents of Shitamachi is contrary in all respects; they maintain the emotions and traditions of the common people of the Edo era: they are fond of traditional public entertainment; they are frank, amiable, religious, and cherish good relations with their neighbours; and they take pride in hard, honest work. They love to celebrate and have a bit of the rebel in them. The true-born Edoite represented the ideal type. It is a type reminiscent perhaps of the one corresponding to Évian-les-Bains in France or Ibn al-Balad in Cairo.[21]

The culture of the common people has been kept alive in the so-called town association. It is similar to Cairo's *harra* of the past or to the *mahalla* in other Muslim cities, but it differs in the following points. The Japanese town association is an autonomous neighbourhood organization for mutual assistance and friendship in which members are all heads of families in a specified block. It has as its subsidiaries organizations of young men and women, children, and old men and women, categorized by age and occasionally by sex. It used to carry out such functions as fire-fighting, crime prevention, and nightwatch. In its function as the terminal unit of urban administration, it is a unique feature of the Japanese city (Nakamura 1979).

There are many theories regarding the origin, functions, and organization of the town association. Tokyo, for example, underwent the upheavals of the Meiji Restoration, the great Kanto Earthquake of 1923, the damages of World War II, and the post-war expansion. And accordingly, the town associations within Tokyo differ from area to area. Reflecting the diverse social characteristics of the inhabitants of the city's areas or blocks, the attitudes of people toward their neighbours and the manner and extent of their participation in the town associations also differ.

In particular, liberal intellectuals of the middle class still remember the dismal experiences in the years of the military fascist regime, when the town association was used for economic control and spying on neighbours. Some resent the present-day town association, seeing it as a vote-collecting machine to support the party in power.[22]

A Provincial City Case-Study:
Traditional Technology in Kanazawa

The Tokyo metropolitan area forms the primate city, and, on the prefectural level, the seat of the prefectural capital is considered the regional primate. In Ishikawa Prefecture, for example, with a population of 1,110,000, Kanazawa City, the prefectural capital, has 410,000 residents, while the second largest community, Komatsu, has only 100,000, and the third largest less than 70,000 (as of 1984).

Kanazawa was the fifth largest city in Japan in the Edo period, after Edo, Osaka, Kyoto, and Nagoya. It now ranks thirtieth of 652 cities. Kanazawa has preserved the unique culture of an old castle town. Except for the Tokugawas, no other feudal lord had an income as large as the lord of Kanazawa, who was a member of the Maeda family. That the site of this lord's Edo residence is now the campus of Tokyo University gives an indication of the power he held.

The Maeda family developed Kanazawa into the largest centre of traditional arts and crafts after Kyoto. The climate in Kanazawa is the most suitable in Japan for the production of lacquer ware, in which the preliminary production process is woodworking. The technology of woodworking has a close relation with the technology for manufacturing metal tools and cutlery. The pinnacle of this technology was represented by the techniques of forging and heat treating to produce the blades of Japanese swords. Heat-treating technology was developed with the aid of high-quality charcoal. The local supply of good firewood and of high-quality charcoal was indispensable to the production of the internationally famous Kutani pottery and of metalwork. The technologies for Japanese paper, fish-nets, fine linen (*habutae*), silk fabrics (pongee), japanned works, and dyeing (*yuzen*) had also been developed, and their products were high-quality Kanazawa specialties that had markets throughout Japan. Other famous traditional fine arts of high value had evolved as products of high-level technology, but they were made only in small quantities for limited markets.

Representative of these was damascene. Damascene refers to inlays of thin gold or silver lines into armour and arms. The secret of Kanazawa damascene was that, regardless of the design, the bottom of the inlay was wider than the surface, thus displaying practical characteristics in its resistance to shock and in its elasticity. Because of the high artistic and practical value of Kanazawa damascene, masters of goldsmithing (including gold-leaf artisans) were accorded the status of samurai.

There were 45 such masters on the eve of Meiji, and each had more than 10 craftsmen under him and an equal or greater number of apprentices. Including the craftsmen employed for intermediate processing, the people engaged in related sectors of technology and services, suppliers of raw materials, and the family members of all, the total number reached to several thousand. When other types of craftsmen and their supporting sectors are included, the total accounted for nearly half the total population of Kanazawa (which was less than 100,000). The political changes brought on by the Meiji Restoration, however, quickly deprived them of their steady customers and protection by the clan, causing widespread unemployment and poverty.

The new central government's general policy of promoting industry, however, also provided an opportunity to the people in Kanazawa. The new mayor, a former low-ranking samurai, made plans to industrialize the traditional arts and crafts. He campaigned among influential and wealthy citizens to initiate a copper-ware company. All of this represented a switch from government patronage to civilian needs, from inlaying on iron to inlaying on bronze and red copper, and thus a major conversion in materials and technology. It also meant a transformation of craftsmen into small entrepreneurs and skilled workers.

Kanazawa's participation at the request of the Japanese government in the international exhibition in Vienna in 1873 provided the necessary momentum that advanced such technological change (Tanaka 1980). Later, Kanazawa goldsmiths were given a boost of confidence after winning a prize at the First All-Japan Exhibition of 1877, held to promote greater awareness and exchange of regional technologies. This positive trend continued with the international exhibition in Paris of the following year.

In the mean time, the Ishikawa Prefectural Industrial Museum was established in 1876 through the efforts of pioneering local government officials, and vocational schools were opened with the object of enhancing and modernizing the traditional local craft industries.

Contrary to the practice of passing down in secret technology from the master to his apprentices within each workshop, the open technical training in the vocational schools contributed to forging new leaders in the local traditional crafts industry. As the technological level rose, the educational system became more refined. Thus were laid the foundations for the eventual establishment of Kanazawa Technical College and, after World War II, Kanazawa

University of Fine Arts and the Department of Engineering of Kanazawa University (Tanaka 1980; Koyano 1979).

Although the process of modernization and improvement of traditional technology appears at first to have evolved smoothly, it was, in fact, challenged by some major bottle-necks, ones that were affecting the entire nation. Among them were the Sino-Japanese War, the Russo-Japanese War, World War I, and—the most important—the rice riots of 1918.

Urbanization meant an increased demand for rice. The steep price rises caused by wartime inflation severely affected rice (the price more than tripled in two years), which was the staple food for people living in cities. The rice riots were ignited by a protest movement of fishermen's wives in a small port town on the Japan Sea coast. They attempted to prevent the locally harvested rice from being delivered to other prefectures. They attacked the rice retailers, forcing them to open their warehouses and sell their rice cheaply. The riots spread quickly throughout the country, not only to big cities but also to factory and mining towns and even to rural villages. The exploitation by landlords was so severe that 90 per cent of the peasants producing rice were too poor to supply themselves with rice. For three months in 60 cities, towns, and villages, there occurred the biggest riots since those during the Meiji Restoration, disturbances that could not be calmed without the use of military power.

Rapid industrialization had eradicated the built-in stabilizers of premodern society, while the functions for adjustment modern society should have provided had not yet matured. The rice riots were an eruption of class conflict caused by this situation.

Another form of the class problem, tenancy disputes, had spread throughout the country, and reached a peak in the 1910s. The disputes involved tenant farmers who were protesting against the conditions of extreme poverty and the lack of rights brought about by a rigidly structured, parasitic absentee landlord system. The protests eventually led the government to adopt legislative measures for the protection of tenancy rights.

It is noteworthy that the rice riots occurred at the stage of the formation of a national technology network. Those at the vanguard of the rice riots in Kanazawa were the craftsmen, especially the gold-beaters (Hashimoto 1980). During this time factory workers were not yet sufficiently organized as a group in Kanazawa, as the dual structure of the low-income artisan class and what we're calling the pauper class was still in existence there. It was perhaps natural, then, that the craftsmen were the main leaders of the protest campaign.

It is said that the rice rebellion in Kanazawa was well organized and highly disciplined. This is probably because the main body of the rioters comprised craftsmen of traditional arts and technology who were socially and culturally conservative and who adhered to a strict aesthetic code.

The radicalization of even such conservative craftsmen as these, who devoted their lives to their aesthetic pursuit, was the product of a structural change in society that was quite different from the Meiji Restoration. For even though the craftsmen had successfully broken away from their small local market to secure a nation-wide market for their traditional technical arts, their standard of living was not thus raised, but in fact it began to drop as a result of the steep price increases.

Kanazawa, though ranked high among other cities when agricultural production was the mainstay, fell in position as the nation industrialized. This was because industry in Japan has tended to develop along the Pacific coast. However, with attention turned in recent years to regional development, Kanazawa—heretofore free of industrial pollution—now has a new and rich potential.

6

Agricultural Technology and Development

Development and Agriculture

Technology transfer is most difficult in agriculture because of the differences in natural conditions, such as weather, geographical features, plant ecology, and irrigation, which overlap social and institutional restrictions.

When an agricultural technology is stable as a result of the limitations imposed by the existing national conditions and social system, the limits of production are empirically foreseeable. The size of the population that can be supported and the standard of living that people can afford are also thus defined. At this stage, agricultural technology forms a strong, independent system, that is stable and fixed and shows a resistance to new technology and the instability inevitable in a new technology.

If a population explosion occurs, the agricultural sector is forced to attempt to cope with the greater demand, but because of the fixed and independent nature of agricultural technology, a serious difficulty arises. The response by agricultural technology is either (1) to maintain the technology at the same level, aiming at extensive development, that is, enlargement of the area farmed, or (2) to realize spontaneous development in technology and productivity. Only in lucky situations are both possible.

However, in the case of a rapid population increase, such as a doubling in 20 years, the surplus required for either 1 or 2 will have been exhausted. Furthermore, the existing materials and facilities will have become worn from excessive use. Here lies both the fundamental difficulty of development and the urgent necessity for it.

Viewing the current situation in developing countries, we can see that the population explosion in Japan beginning at the end of the eighteenth century and the prevailing technological conditions were identical to those to be found now in the developing countries. In Japan also, rapid and abnormal expansion of urban population occurred, and as the great political changes of

82

the Meiji Restoration paralysed the functions of the existing system, the damages from natural disasters became serious and widespread. This caused a rapid increase in the number of uprooted peasants, and, as previously pointed out, many of them moved into the cities—to form new classes of urban poor. However, because the population tripled over a period of 100 years, the increase was not so explosive as in today's developing countries.

Japan found solutions in industrialization and raising agricultural productivity. One result of this, however, has been a loss of self-sufficiency in food, except for rice. In terms of calories, Japan has only 50 to 60 per cent self-sufficiency, a level similar to that in West Germany. A greater dependence on industry and specialization largely account for this.

Formerly in Japanese agriculture double cropping was common, and, at the initial stage of industrialization, Japan's agriculture sector was able to satisfy industry's demand for raw materials. But nowadays, farmers specialize in rice, and cultivation is carried out on a much larger scale. As a result, agriculture now depends on industry for fertilizers, agricultural chemicals, machinery, and fuels. Indeed, agriculture has deepened its dependence on the industrial sector.

Recently a nutritionist made the noteworthy observation that one reason Japanese agriculture rushed from the double-cropping system to a rice monoculture is that the calorie per unit production area is much higher (about two times) for rice than for wheat or potatoes, besides the traditional preference for rice. And, while rice requires an input of labour three times what is required for wheat, its employment absorption capacity is higher. In other words, the emphasis is on productivity of the land rather than on productivity of labour. This means, however, that the mode of rice production continually absorbed the growing labour force resulting from a rapidly expanding population and, simultaneously, pushed agricultural labourers close to the point of starvation.

The successful transition to rice cultivation illustrates the adaptability of Japanese agriculture and farmers in the face of changing needs and demands. One reason for this was that in Japan even the non-landholding tenant farmers were free to manage the fields they were renting—free to decide what to grow, how and when—unlike in the developing countries of today. In addition, as a result of having been monoculturized, Japanese agriculture has come to resemble colonial agriculture. The difference, however, is that Japan has also acquired industrial power.

With the aid of fertilizers and agricultural chemicals, Japanese farmers yield the greatest volume of rice per unit of area in the world. However, the use of chemicals, especially over the past 20 years, has caused the soil to lose rapidly the fertility that had been built up over more than 100 years by the use of organic fertilizers. The loss of fertility has necessitated additional input of chemical fertilizers. Input per hectare in Japan (372 kg) is the third highest in the world, following the Netherlands (789 kg) and West Germany (471 kg). When crop type is taken into consideration, Japan is the highest among countries cultivating rice, that is, 2.4 times greater than in China, 5.9 times

Indonesia, 12 times India, and 22.9 times Thailand (according to a 1980–1981 FAO survey). It should be noted, however, that these data do not account for individually supplied fertilizers, so that a simple comparison may not be valid.

A rising reliance on new fertilizers and agricultural chemicals, however, is not without harmful side effects; indeed, ecological changes will be produced that give rise to a situation like that described in Rachel Carson's *Silent Spring*. In the early 1970s, it was reported in Japan that 40 farmers die every year from poisoning by agricultural chemicals. It is said that chemical fertilizers have made the soil so hard that it cannot be turned except by tractor. While countermeasures are being applied, this is not a situation that can be quickly resolved.

The most promising measure is the recommendation for a breakaway from a rice monoculture to a diversification of crops. It is argued that, just as modern industrial technology has developed by virtue of converting the mass production of one commodity to the production in small quantities of various commodities, agriculture has no other course than to change from rice cultivation to the cultivation of different crops. Even though a market has been secured for farmers in which they are able to sell their rice at the world's highest prices, their relative income has continuously fallen and cannot support them.

This latter is one reason for the increase in category-2 rice farmers—that is, farmers with side jobs (who earn bigger incomes from the side jobs than from rice cropping). On the other hand, however, the average age of farmers is too high to expect an immense volume of labour input required by the change to crop diversification, and there are few among the younger generation willing to undertake the intensity and long hours demanded by agriculture.

The Change in Agricultural Technology, from Emphasis on Land Productivity to Emphasis on Labour Productivity

Pre-Meiji Agricultural Methods

During the more than 200 years preceding the Meiji Restoration, the Tokugawa shogunate promoted agricultural development in various ways, and it also exercised strict control over the use of natural resources. Despite the abuses that come with a long-term, stabilized political power, in general, the Tokugawa government was remarkably diligent in working to preserve the environment. This was because its material basis was in agriculture, which, of course, depends on nature. Thus, its policies of forestry conservation and river improvement constituted an important basis of the rice-culture economy.

In the shogunal domain of Kiso, famous for its Japanese cypress forests, cutting trees was strictly controlled by the administration. This kind of severe control was exercised not only in Kiso but also in the domains of other clans, because the forest was regarded both as a supply of wood and as a major reservoir of water, essential for agriculture. Cutting small trees was allowed because it was believed to benefit the larger ones; the cycle of cutting and regrowth was wisely maintained.

In the Seto area, control over the forests (and forest resources) was relaxed after the Meiji Restoration and after the production of pottery for export had increased. As a result, the reckless exploitation of the forest resources began. The devastating landslides and floods that came as a result, however, eventually brought the exploitation to a halt. Attempts at the autonomous administration of basic natural resources, that is, plans for forest reserves, appeared, and progress was made in energy conversion (in the beginning, this took the form of a change in the place of supplies and then in the materials themselves). During the Edo period, the development of salt farms required the felling of trees in the domains of some clans along the Seto Inland Sea. For purposes of conservation and river improvement, however, it was necessary for these clans to reforest these areas.

Many reformers of agricultural technology appeared in Japan in the nineteenth century, and a number of agricultural handbooks became popular.[23] It is noteworthy that most of these books were concerned with the technology of civil engineering and, in relation to treating the water problem (construction of irrigation-fed paddies and dry-rice fields reliant on surface or underground drainage), urged crop diversification.

These practical reformers, or agricultural "wisemen," appeared on the scene just when the modern population explosion started. The crisis in agriculture and in the peasant household created a need for the wisdom and rich experience of these farmers, and in fact, their technology played an important role in increasing production.[24]

After the Meiji Restoration, new agricultural technology was introduced into Japan from Europe. Agricultural colleges were established in Tokyo and Sapporo. As Sako Tsuneaki, a student in one of the schools of those days, said: "What we learned was not Japanese agriculture but British agriculture." Tamatsukuri Yoshizo, another student, said, "We had no knowledge at all of Japanese agriculture." And it seems what was true for the students was true for the teachers (all of whom were foreigners). The innovative agricultural scientist Yokoi Tokiyoshi (1860–1927) was a later student of one of these agricultural schools. At the turn of the century, however, modern Japanese agricultural science had not yet been created.

Introduction of Horse Tillage and the Reorganization of Farm Lands

The development of the agricultural revolution through the efforts of the agricultural reformers of Meiji was ahead of the industrial revolution in Japan, much as in England, where the Norfolk agricultural system developed before the Industrial Revolution. The core of British agriculture was stock-raising and the cultivation of wheat. The technological system and conditions of British agriculture were quite different from agriculture in Japan, where horses or cattle were used as work animals in the cultivation of rice, in which irrigation was the decisive factor.

However, the two did share some common characteristics. In the Norfolk system, scattered, small farm lands were integrated within enclosures, and a system of rotation of crops for cereals, pasture, and root-crop feed was adopted. An important element of this technological change was deep tillage by horse-drawn plough. Deep tillage using draught horses was developed in Kyushu (especially in Fukuoka) also as a new agricultural method. The Japanese counterpart of the British enclosure was a rearrangement of agricultural lands by which land that had been divided into small pieces was repartitioned or integrated into a larger unit. Irregular paddies were thus transformed into square or rectangular shapes with one acre as the basic unit. Odd-shaped fields disappeared, and planting in check-rows increased yield.

The rearrangement of farm lands and a reconstruction of irrigation and drainage ditches and straightening of farm roads (which increased transportation efficiency) constituted a fundamental reform of the system and contributed to greater efficiency of irrigation-based farming. As such, the reform required a consensus among all interested parties and also capable leaders co-ordinating the various interests. A well-planned reorganization of agricultural lands not only increased efficiency but also often meant an increase in total acreage. Even when progress was modest, it was of incomparable value for the peasants.

One technological characteristic of the Norfolk system was intertillage by means of drill and plow. Intertillage is a common component of agriculture in humid areas, and it has been taken for granted in Japan, while in the West, contrarily, it was more commonly associated with gardening than with agriculture.

The wheel-less plough in use in Japan was comparable to the Norfolk tiller.[25] This plough was an upright holding plough and its operation required skill and strength; however, because it made deep tillage possible and raised production, "ploughing instructors" were in demand throughout the country. Many improvements were made to it until eventually a new type of short-bottomed plough was developed that eliminated the heavy work required in operating the earlier wheel-less plough.

Then, at the beginning of this century, a short-bottomed plough with a head that could swing both right and left was invented; unlike its predecessor, this new swing-type plough made it possible to turn the soil in the same direction both coming and going. The plough was well suited to flat-land tillage and represented an epoch-making improvement in Meiji agriculture.

A basic difference between Meiji agricultural methods and Norfolk-type Western agriculture was that the latter required a large capital investment and relied on the economic advantage of labour-saving, large-scale operation, while Meiji agriculture was characterized by small, mixed-crop management. Since pre-Meiji times, a high input of labour was accepted as natural, and cultivation and the application of home-produced fertilizer using family labour was the norm. This system inevitably led to small-scale farming and multiple cropping. The introduction of tillage innovations in Fukuoka, however, made possible a yield of as high as three times the national average. Once horse-plough tillage spread all over Japan, agricultural productivity was raised by 50 per cent.

The Fukuoka method was suitable only for dry fields, however, and so it could not be used in the wetland areas of eastern Japan. A huge investment in drainage facilities was necessary to create dry fields.

The Meiji government employed the Dutch-type low-water engineering method for agricultural development, especially for irrigation facilities, and while it was difficult to apply to the drying of fields and thus posed somewhat of a technological bottle-neck for Meiji agriculture, the government adopted the method because it accommodated the water-borne transportation system that existed at the time. Until the 1910s, when the railways became popular and a transportation system combining horses and the railway was established, water-borne transportation was vital for the transportation of heavy cargo, and, in general, was responsible for meeting more than 50 per cent of all transportation needs in Japan.

Flood damage caused by the low-water engineering system is said to have occurred every three years. But, as the flooding also provided needed enrichment to the soil, it didn't cause the farmers serious concern. Nevertheless, productivity was low and unstable, and measures were necessary to raise productivity and maintain it at a higher level. One such measure was the prevention of floods through high-water engineering, which was combined with a modernization of irrigation and drainage facilities. With the big flood near the turn of the century (1898) as a turning point, high-water engineering and, at the same time, irrigation using the major rivers gained widespread currency.

Establishment of the Modern Irrigation System

From the 1910s to the 1920s, notable improvements were made in water-supply facilities, in waterways and water distribution (to individual fields). Full-scale development started in 1923 with a government "project for the improvement of arterial waterways for irrigation and drainage" that stretched across several perfectures. This meant that, from around this time, the traditional small-scale irrigation integral to Meiji agriculture would undergo a complete transformation. Meanwhile, the adoption of high-water engineering corresponded to the spread into agricultural villages of the railway and other replacements of water-borne transporation.

Another important change occurred around this time; namely, a shift in agricultural investment initiative and activity from the large landlord class to that of small and medium landlords and landlord-farmers.

The price of rice, which had been on the rise since the beginning of the Meiji period, tended to fall in the 1910s as good-quality rice began to be produced in large quantities and supplied on a regular basis. This did not mean, however, that Japan had attained self-sufficiency in rice or that the peasants could have rice three times a day. There were more than 4.5 million households of tenant peasants—more than 50 per cent of the agricultural population—and they had to pay an average of 55 per cent of the yield to their landlords for rent. Scholars have defined the relation between landlord and tenant at this time as "semi-feudalistic." Exploitation was harsh, and the population explosion intensified the competition for tenant farm lands, which made it possible for owners to maintain high-level rents.

Around 1890, one per cent, or approximately 40,000, of the agricultural population in Japan consisted of big landlords with more than 10 hectares of land each; 470,000 middle-sized landlords with 2 to 10 hectares; and 3.5 million landlords holding smaller parcels of land. From around the turn of the century, the big landlords and the wealthier middle-sized landlords became absentee landlords. As the big landlords became less interested in investment to improve the land, the middle-sized and small landlords, especially the landlord-farmers, began to initiate their own projects. This class of farmers did not have the financial capacity, however, to bear the expenses of constructing irrigation and drainage waterways off the major rivers, so it was necessary to rely on government aid.

The government adopted a formula in which the farmers first co-ordinated the interests of the concerned villages and towns, after which it reviewed a project's technological feasibility before accepting or rejecting it. The system encouraged the farmers to take the initiative, while the government avoided direct involvement in the difficult matter of the engineering aspects of the projects, leaving it to the farmers. The leaders of the agricultural associations involved in irrigation or land improvement projects were not necessarily from the influential landlord families. They had a thorough knowledge of the customs of each farmhouse regarding the use of farm land and irrigation water and a great talent for mediation and documentation required in negotiations with the government.[26]

Such strong leaders could be found in many areas of the country, but not all villages and towns were so blessed. And the differences resulting from their presence or absence are apparent even today (Hatate 1981).

The Emergence of Tenant Problems

Poverty and frequent tenancy disputes characterized Japanese agriculture in the 1910s and 1920s. The disputes began sporadically immediately after the Russo-Japanese War and increased rapidly in the 1920s. As a general tendency, they spread gradually from western Japan, an agriculturally advanced area, to eastern Japan, and after the depression of 1929, to Tohoku and Hokkaido. In response to the tenancy disputes (and also the labour disputes) the government intervened and brought harsh oppression to tenants. Eventually however, because it was obvious that increased agricultural production depended on tenant labour, the tenancy disputes were successful in bringing down the rents.

Many poor peasants were compelled to send their young daughters to work in cotton-spinning factories, receiving a small amount of advance money on indentures. The peasants' poverty aroused a strong sense of crisis among young military officers. As healthy young men were conscripted from poverty-stricken families, the militarists became seriously concerned about a further deterioration in living standards of peasant families, who were contributing their most important labour resource as soldiers for the state.

The military fascists urged the reform of "landlordism." As far as a perceived need for social reform was concerned, the right-wing and left-wing radicals shared the same understanding of the issues; the nature of the proposed reforms, however, differed greatly between the two. Whether reform in the name of "national polity" (i.e. the Emperor) or by workers under the name of socialism, the need for social justice was clear to both.

The rice riots of 1918 were the first eruption of the problem. A series of spontaneous urban riots continued across Japan until the autumn harvesting, bringing to light the serious structural problems (e.g., of allotment, production, and distribution) that post-Meiji modernization and industrialization had failed to remedy.

But Meiji-Taisho Japan was unable to solve these problems, and they were taken over by the "changeling" military fascists, who displayed their incompetence to cope with the problems in their rush to war and defeat. It was only after World War II that land reform was realized, as part of the "democratization of the economy," that Japanese agriculture finally saw a real turning point.

Under the restrictions of the landlord system and within the framework of Meiji agriculture, improvements in agriculture were inevitably confined to the development of technology that would effect improvements in water utilization and in small-scale farming tools. The turning point in agriculture after World War II came with a shift in emphasis from land improvement, i.e. increasing land productivity, to raising labour productivity through more efficient use of water resources and mechanization. Productivity policy, which was tied to the establishment of the landed farmer that the tenancy disputes helped to realize, shifted its priorities from the land to the farmer.

Electrification in Agricultural Villages and Water Utilization

In relation to electrification, there were some important technological changes that occurred. First, we must mention a small pump that gained wide use in the 1910s. This pump was used to raise water from lower-level irrigation channels and meant a great savings in labour. It was also used to drain water from paddy fields. The diffusion of the pump paralleled the electrification of agricultural villages, as it used surplus electricity generated by the hydro-generating system that had already been established. By the 1920s, the change-over from the engine-type to the electric-motor pump was progressing smoothly. The important point is that the widespread availability of this low-cost pump was made possible by industrial development.

Second, as the use of electricity in agricultural villages progressed, threshing and rice-milling machines became electrified, and, of particular importance, electricity was introduced into irrigation. In such areas as the Saga plain, where the level of water for irrigation was low, creek irrigation had been adopted, and electricity greatly simplified operation of the water-gate and liberated peasants from the grueling labour of pumping water by means of a pedal-powered water-wheel, which on its best day could supply no more than enough water for 1.5 hectares (Jinnouchi 1981).

Mechanization

The greatest change in post-war agricultural technology was mechanization through the rapid spread of the tractor and combine beginning in the 1970s. Mechanization dramatically increased productivity, and because mechanization in agriculture paralleled mechanization generally, fuel and repair parts were consistently available, thus eliminating many operational and maintenance problems.

This was in remarkable contrast to the failure encountered in attempting to introduce North American type large-scale (dry-field) farming into Hokkaido at its initial stage of development, a failure in part resulting from troubles with imported machines and lack of servicing capabilities. Before the machines were introduced, a rearrangement of farm lands—much like that when the horse plough for deep tillage came into use—was necessary, this time with a maximum unit of more than a hectare. Agricultural roads were expanded and reinforced, and because this was the second such reorganization, the difficulty was not so great as the first time.

The modernization, however, produced an unexpected change in the villages: a decrease of full-time farmers and a rapid increase of farmers with secondary jobs. Full-time farmers occupied 50 per cent of rural households in 1950, a rate that decreased to 21.5 per cent in 1965 and to 13.1 per cent in 1982. At the same time, the number of farm households decreased during this period from 3,080,000 to 1,220,000 and to 600,000, respectively; that is, by more than 80 per cent in 30 years.

On the other hand, farmers with sideline jobs whose major income was from agriculture occupied 28.4, 36.7, and 16.9 per cent of rural households in the specified years; the percentages for farmers whose income from agriculture was secondary increased sharply during the period, from 21.6 per cent to 41.8 per cent, and finally to 70 per cent.

Although the greater number of independent farmers resulting from the land reform contributed to the increase, it was mainly the result of mechanization introduced into small farming, where holdings of less than 0.5 hectares occupied 41.4 per cent, 0.5 – 1 hectares 28.2 per cent, and 1 – 2 hectares 21.2 per cent. Rational adjustment between the scale of operation and mechanization is usual in large-scale operations, but in Japan, farms larger than 5 hectares make up only 15 per cent of the total. This is the fundamental difference from the EC-type operation, under which the upper 20 per cent of farmers yield 80 per cent of the total agricultural produce.

Consequently, a tendency has developed among full-time farmers who have mechanized equipment to subcontract or lease small parcels of land, for even small, scattered plots, when combined, add up to a substantial effect. The increase in part-time farmers since 1965 confirms this; that is, only farmers who thus accumulated extensive holdings could survive as full-time farmers. Furthermore, the shortage of land brought on by the urbanization that occurred during the period of high growth and super-full employment pushed land prices up, which, while raising the value of farm lands, also had non-farming urban commuters living alongside farm residents.

The enlargement of cultivation by full-time farmers under the subcontracting system is considered by some to be approaching its limit. They reason that because subcontracted lands are small and scattered, much time is needed to deliver machines and farming efficiency is deteriorating. If a return by farmers to a composite farming system is unfeasible, mechanization faces either stagnation or the necessity for a new reorganization of farm lands. Thus, we have arrived at a turning point. According to an early 1980s FAO report, the agricultural population per 100 hectares in Japan is 128, versus 98 in India, 1 in the United States, and 8 in Great Britain. Because agriculture in Japan is different in its basic conditions and crops from agriculture in the West, however, this kind of comparison is of little significance.

Rice cropping requires the input of highly intensive labour, even when mechanized. In terms of calories per production unit, however, rice is more advantageous than wheat. This is one reason why, after rejecting the recommendation to government authorities by American and European specialists ("development advisers"), farmers who settled in Hokkaido persisted in growing rice and made Hokkaido the biggest rice-producing area in Japan. They discovered they could get enough sunshine to grow rice despite Hokkaido's high latitude. Here we see the importance of not depending too much on foreign specialists.

With reference to the current situation in the developing countries, note that even though the central and local governments in Japan have made huge investments in irrigation facilities, the basis for agricultural development, their administration and maintenance, based in the villages, have been left in the hands of existing institutions. This has contributed significantly to the successful development of irrigation facilities in Japan, compared with the situations in India and Pakistan, for example. We are not "participation romanticists"; the important point is that the uselessness of the uniformism of bureaucratic control has been most readily observed in agricultural technology, in its administration and maintenance.

I wish to add that too often the role of agricultural experiment stations, which were established throughout Japan after the Meiji period, and the co-operative relations between these stations and local farmers have not been properly examined. We attach importance to the agricultural experiment stations because they have been able to respond to and meet local needs. Especially noteworthy have been the improvement in plant and horse breeds (in which the army was also engaged) and the subsequent improvement and diffusion of cultivation techniques. The task of raising agricultural technology to a mid-level in developing countries is urgent and, consequently, we are greatly interested in these stations. If there is sufficient interest, perhaps such experimental stations could be the subject of an international comparative study.

The agricultural schools have also played a crucial role. Most leaders in agriculture, including those of the irrigation associations, were graduates of agricultural schools. Many of the young leaders in agriculture today are university graduates, and their roles are still significant.[27]

The co-operative associations in Japan have also played an important part. Their activities pertaining to the supply of credit or financing have been treated in our sub-project on Technology Transfer and the Role of Financial Institutions.

A record by Namie Ken (1981), a pioneer in establishing libraries in agricultural villages, has been included in our project; it is a historical document devoted to Namie's work with farmers in a suburb of Tokyo. Namie was regarded by the military fascist regime of the 1930s as a dangerous political activist.

A study that does not mention fishery in relation to the population explosion and the food problem may be considered inadequate as a study of the Japanese experience. Nevertheless, within our limitations we had to adopt preferential policies in choosing our topics, which, we realize, opens the door to criticism, but which also ensures the unlimited possibility (ideally) for the kind of methodological exchange proposed by the United Nations University.

Very briefly, however, it is important to note that, with the increasing dependence of agriculture on industry (for fertilizers, pesticides, etc., for the development of agricultural productivity), and the resulting pollution, the fishery sector has begun to suffer great damage, which has brought these two sectors into intense rivalry.

7

Transfer and Self-Reliance in Iron and Steel Technology

Pre-conditions

The technology to manufacture iron and steel is the most representative of the technologies that were once imported and that are now being exported. Iron-manufacturing technology is a complex, large-scale technology that is not easy to transfer. Thus, approximately 40 years were required before independence in this technology was attained and, indeed, 100 years before technological exports could begin, after additional transfers had been made following World War II. The urgency of development today, however, does not allow such slowness. The only option then is to identify the most feasible measures from other countries' development experiences through field-work and dialogue. Following are a few suggestions regarding the technology choices to be made.

Unlike most developing countries today, Japan was in an advantageous position concerning the pre-conditions required for iron-manufacturing technology. Nevertheless, the technology that was transplanted to Japan was not the most advanced available at the time; it was merely the most widely used.

Japan's first reverberating furnace was constructed in the Saga domain in 1850 to forge big guns. Its technological level was pre–Industrial Revolution, and it was operated using charcoal, water-wheels, and cold blast. Since the reverberator converted pig-iron to malleable iron (wrought iron), obviously the supply of pig-iron had to be ensured. The pig-iron produced by traditional Japanese technology was not of a quality high enough, however, to be used for making big guns. Accordingly, unless the iron was produced using Western methods, the reverberator was useless. So it became necessary to construct a blast-furnace. This example demonstrates the need, when transferring technology at a point downstream, to go to the upstream source; the nature of technology is connective and cumulative.

Thus, in 1854, the Satsuma clan constructed the first blast-furnace in

Japan. While steam-power and coke were being used in blast-furnaces in Europe, in Satsuma, the furnace was designed to operate using a water-wheel and charcoal because of the scarcity of coal. Note that the furnace was designed to conform to the existing limitations. Like its predecessor, this blast-furnace was limited and could not be developed into a modern, full-scale industrial technology. Nevertheless, because the furnace rationally corresponded to the size of the market for the munitions industry the clan operated, its design and scale were appropriate.

A series of difficult problems relating to raw materials and ventilation marked these early attempts at iron manufacturing. This comes as no surprise when one learns that the Saga clan's reverberator and the Satsuma's blast-furnace were built using a single technical guide, Ulrich Haguenin's *Het Giet-wezen in s'Rijks Ijzer-Geschutgieterij* (te Luik, 1826), without access to any models or instructors. Any similar attempt made today would be censured for its recklessness. Nevertheless, the projects achieved a certain, if limited, degree of success. The success meant that Japan had attained a level of technology comparable to that of Europe on the eve of the Industrial Revolution. At the same time, on a practical level, it became clear that the sand being used as the raw material for pig-iron in Japan was unsuitable for casting.

Although the technological level before the Meiji period was high, the gap between it and modern technology, including technology for developing raw ores, was great. The Meiji Restoration broke the shogunate's monopoly on the importation of technology and created circumstances in which technological guidance could be obtained directly from foreign engineers.

Failure and Recovery of the Kamaishi Ironworks

In its seventh year in power, 1874, the new Meiji government initiated construction of a modern ironworks at Kamaishi mine in present-day Iwate Prefecture. The orders for a plan were given to L. Bianchie, a German employed by the government, and Oshima Takato (1826–1901). Oshima, whose father was a medical doctor in the Nambu domain, studied medical science at Nagasaki and later concentrated on gunnery, mining engineering, and refining technology. He constructed a reverberating furnace and a Western-style blast-furnace at the Kamaishi mine in the Mito domain. Oshima was the translator of the Haguenin work, and as a member of the Iwakura Mission to Europe and America, he had an opportunity to further his knowledge of the West's technology.

The two proposals submitted by Oshima and Bianchie differed in basic design and location. Bianchie proposed a design for two large and highly efficient blast-furnaces (25-ton daily output) and a railway to transport the iron-ore; he even drew up plans for puddling and rolling.

Oshima's proposal was for five small furnaces (5 to 6 tons daily) and a horse-drawn tramcar for transportation. The Nambu domain was famous for its cast-iron goods, good-quality charcoal, and sturdy horses. Oshima's plan was a capital-saving design well suited to the technological conditions present in the Tohoku area at that time and thus appropriate for starting up an enterprise (Iida 1979). Perhaps the lesson here is "start small and grow big."

Nevertheless, Oshima's plan was not adopted by the Ministry of Works, and he was posted to the Kosaka mine in Akita Prefecture, obviously a demotion.

The government-operated Kamaishi Ironworks imported not only the blast-furnace, air-heating furnace, and machines for the puddling plant, but also the locomotives, freight cars, and rails (and presumably the ties, which were made of iron) for the railway system that connected the port of Kamaishi with the mine and the place where the charcoal was produced. For engineering, construction, and operation, the government employed British engineers and foremen.

The works started operation in 1880, seven years after the plan was drafted. After 97 days of smooth running, operation of the blast-furnace ceased because of a fire in the charcoal-making shop that caused serious damage, necessitating that the fire-bricks lining the inner walls of the furnace be replaced, and a shortage of charcoal.

Firing resumed after more than a year's disruption, in February 1882. Because of the charcoal shortage, coke was used, but the coke's inferior quality brought on another stoppage in mid-September, after only 196 days of operation. This second failure, so soon after the first, compelled the government to close the works in December.

Although ostensibly the failure at Kamaishi was due to a shortage of charcoal, the area was famous for its production of good-quality charcoal. One must conclude, therefore, that other difficulties, for example problems with transportation and the procurement of raw materials, were also instrumental in the decision to shut down.

In other words, there were M_1 and M_4 difficulties. But was that all? If Oshima's plan, which had the advantage of diversification of risks, had been adopted, the failure could likely have been avoided. One can discern in the government's attitude and actions an excessive reliance on foreign engineers and at the same time a contempt for its own.

The government sold the remaining materials (charcoal and iron-ore) to Tanaka Chobei, a purveyor for the government. After repeated repairs to the blast-furnace, he succeeded, on his forty-eighth attempt, in producing iron. Tanaka's modifications of production and facilities resulted in an operation less like Bianchie's plan and more like Oshima's original proposal, and this undoubtedly contributed greatly to his eventual success. Foreign engineers and foremen were no longer in attendance, and the workmen were using ores that had been rejected by the previous management because of their "inferior quality." Making use of these ores was a key factor in Tanaka's success.

Tanaka has left a record of some of the problems regarding transportation and fuel problems that were encountered—what we're referring to as questions of technological links and support services. It is apparent that he had correctly grasped from the manager's point of view what Oshima the engineer had reasoned. He agreed with Oshima's idea of starting a moderately sized operation and enlarging it gradually.

Tests showed that the iron produced at Kamaishi was equal to the world's best-quality iron, manufactured by Krupp in Germany, and that it was usable for military purposes. Kamaishi was thus able to acquire, for the first time, a stable market. However, the authorities forced Kamaishi to renovate its facilities, update quality control, and improve transporation. Noro Kageyoshi (1854–1923), professor at Tokyo University and a member of the first generation of recipients of a modern engineering education, was appointed as one of the technological advisers. One of the 25-ton blast furnaces was successfully put back into operation, and, for the first time, success was won in the technology of using coke.

Japanese engineers brought the technology of iron manufacturing in Japan into the modern age: In 1894, when the blast-furnace was restarted, the mill produced 13,000 tons, a 50 per cent increase over the previous year, and in its twelfth year, output exceeded total production from all foot-bellows-type iron mills.[28] But foot-bellow mills survived, and are a good example of the toughness of traditional industry.

In 1895, a British type of rolling machine that had been imported and operated at the plant was able to be repaired by workmen at Kamaishi. In addition, rails, plates, round bars, square bars, and flat irons were manufactured with Kamaishi's own pig-iron, though in only a small, 5-ton quantity. Only 40 years had passed since the Saga clan had groped for the technology for a reverberating furnace.

Thanks to good-quality charcoal and other advantageous pre-conditions, in 40 years Japan had caught up with modern iron manufacturing technology, which had a 200-year history. More recently, Korea has, through its efforts, done the same in 20 years.

Although Tanaka's operation eventually corresponded to Bianchie's plan, this is not to say it was in fact an appropriate starting point; Bianchie and the other foreign advisers were mistaken in regard to the scale and links of technology. Production only got under way successfully once the operation had been reduced to a smaller scale. The Kamaishi case clearly demonstrates how important the choice of a rational scale of operation and technological level and the management of technology are for technology transfer. The second lesson to be learned is that final responsibility for solving the problems should be left to the engineers of the importing country.

There are some technology historians who maintain that Kamaishi was technologically successful because it was able to operate for approximately 100 days. We do not support this position, for the simple reason that technological success is determined by the realization of the full potential, the full economic or physical life span of a technology or operation. Any industrial technology must be used to the limit of its physical or economic life span. What determines this is the technology of daily operation, maintenance, and administration. The foreign engineers at Kamaishi failed in the first stage of technology transfer. Their failure highlights the differences in approach between Bianchie and Oshima, between a techno-scientist and an engineer.

Failure at the Yawata Ironworks

In late-nineteenth-century Japan, the consumption of steel and wrought-iron is estimated to have been less than one kilogram per capita. Considering that the capacity of a blast-furnace is more than 1,000 tons a day, an estimated annual per capita consumption of more than 20 to 30 kilograms is necessary for the stable operation of a mill with a blast-furnace system continuously manufacturing steel from pig-iron. If the population is small, obviously a great per capita consumption or foreign markets is required.

Although modern iron manufacturing had been achieved at Kamaishi, capacity remained at about 25 tons per day, and Japan had to continue importing pig-iron and steel. To meet the government's development target, therefore, a project was initiated to construct a modern ironworks for the continuous operation of pig-iron and steel manufacture. The result was the state-operated Yawata Ironworks in Kyushu, western Japan.

The impetus for the ironworks was provided by a snag in the delivery of weapons the government had purchased that occurred during the Sino-Japanese War. The weapons were held up in Singapore due to a diplomatic wrangle, and Japan realized that, in the government's words, "if the war were to be prolonged, it would face a situation of great difficulty in the supply of weapons." The importance and necessity of weapons independence was clear. In pursuing this independence, however, one problem that had to be faced involved standardization, a subject we will discuss in detail later because of its decisive importance for industrialization generally.

The new ironworks were to be managed by the navy, and after one upset, Blast-Furnace No. 1 began operation in February 1901.

The designer was W. F. Luhrman, and the nominal capacity of output of pig-iron was 160 tons. The actual output hovered around 80 tons, and, in addition, the pig-iron was unsuitable for making steel because of its poor quality. The coke consumption was at a deplorable 1.7 tons per ton of pig-iron. (Today, the average ratio is approximately 0.45:1.0; the expected ratio even back then was about 1.0:1.0.) In July 1902, after less than 20 months of operation, the first blast-furnace was shut down.

Considering that Japan's only previous experience was at the Kamaishi Ironworks, whose output was 60,000 tons per year (20,000 tons of open-hearth steel, 4,500 tons of wrought-iron, 500 tons of crucible steel for military use, and 3,500 tons of Bessemer steel for the railways), the Yawata plan was ambitious: It called for the construction of three 60-ton blast-furnaces, two 17-ton Bessemer converters, four 15-ton open-hearth furnaces, six puddling furnaces, and one crucible furnace, besides hydraulic forging and rolling machines. Moreover, once construction started, the original scale was enlarged: planned output was increased from 60,000 tons to 90,000; the blast-furnaces went from three 60-ton units to two 160-ton units; and each of the four open-hearth furnaces from 15-ton units to 25-ton units. The budget was doubled, to the fantastic sum of ¥25 million, part of which would come from the indemnities of the Sino-Japanese War.

As indicated by its English name, the Imperial Japanese Government Steel-Works, Yawata was intended to be a symbol of the nation. Unfortunately, however, it merely repeated the failure at Kamaishi.

To give this some perspective, it might be mentioned that U.S. Steel, established in 1901, had a nominal capacity of 10.6 million tons; thus, even if Yawata's capacity could be raised to 90,000 tons, this was still less than one per cent of U.S. Steel's capacity. It was indeed a tiny mill, although the biggest in Asia.

When the attempt to enlarge the scale of operation failed, Noro, the technical adviser referred to earlier, was recalled. His investigation revealed the following:

1. Design flaws were found in the structures of both the blast- and open-hearth furnaces.
2. The suggested operational procedures were unsuitable for the raw materials available in Japan (the mixture of input materials was inappropriate and the coke being used was of poor quality).
3. There were serious problems with the ventilation facilities.

The production equipment was manufactured by the Gutehoffnungshütte Company of Germany. The company had sent approximately 20 skilled foremen, and the Japanese government hired 3 top engineers for the operation. But the results were disappointing. Besides the design errors mentioned, the operational guidance was poor: the planners repeated exactly the same mistakes as had occurred at Kamaishi in their inclination toward larger scale, the most up-to-date facilities, and blind faith in foreign expertise. Japanese engineers were able, after some effort, to correct the faults, and enough alteration was made to the design of the second blast-furnace under construction to enable it to start normal operations in 1904.

The success was attributable to the efforts of Hattori Susumu (1865–1940), Noro's pupil. The top-level group of engineers had been fired and, except for a foreman for the revolving furnace, had returned to their respective countries at the time of the Russo-Japanese War. They returned home, it seems apparent, because of their loss of confidence.

To give a simple example of some of the difficulties that were encountered, the open-hearth furnace used at Yawata and designed by a German by the name of R. M. Daelen (1843–1905), was characterized by Imaizumi Kaichiro, an associate of Hattori, "as having a most serious defect, the location of the jet, which could be corrected after experimentation, but because of space limitations, it was impossible to improve too short a jet and build a room for residue from the furnace." Imaizumi repaired the furnace, and, in discussing the problem with Daelen directly, is said to have been told by Daelen himself that the design was in fact "a totally untested desk plan."

This represents the sort of obstacles that had to be overcome before technological stabilization was attained.

No piece of equipment or machine can be expected to operate in the beginning at the level for which it was originally designed; newly designed equipment is more unstable. It is the engineer's job to bring the working level up to the intended level of operation. Not uncommonly, operational stability is reached at a lower level than the originally designed output level, and this low-level output often becomes a maximum-output level. Output greater than the design intended should not be attempted because it may cause problems or accidents. If a greater output than intended is attained, it means that technological potential and precision are being sacrificed in excessive concern for safety. Considering, however, that machinery exported to developing countries is used under diverse conditions, perhaps greater importance should be attached to safety, even at the expense of precision.

There are some specialists who argue that the Yawata Ironworks was wisely modelled on German rather than on American plants because the demand structure in Germany, with many types of steel produced in small quantities, was quite similar to the demand structure in Japan. In other words, the technology was carefully selected and transferred. Yawata's failures thus represented a degree of progress compared to those at Kamaishi.

In any event, Japanese engineers corrected the design and operational errors of the foreign experts, and, by 1910, the Yawata mill had gone into the black. Specialists point out that this, together with the introduction of the Solvay coke furnace and the newly acquired skills for making coke, were of great significance.

To conclude this section, it should be pointed out that, although the teachers were poor and the students excellent, the students were still no more than students. They did not have the skills and experience necessary to see a problem on paper or to create a design, and certainly not to implement a design.

From the point of view of the history of Japanese iron manufacture, Kamaishi represents the period of the establishment of technology and its operation, and Yawata the ability to correct and improve and introduce new methods and technologies. In sum, Yawata represented the fourth stage in Japan's technology transfer aimed at self-reliance.

Hoshino Yoshiro, who revised my five-stage theory, claims that the fourth stage, the "ability to design technology," has three substages: (1) complete imitation and additional testing, (2) partial design alteration, and (3) complete renewal of the design. Yawata's second blast-furnace indicates that it had certainly arrived at substage 2.

Iron-manufacturing technology arrives at the fifth and final stage when the technology for treating non-design-type problems has been developed, for example raw materials. Although the conversion to electricity as the power source had begun in the 1910s, the consumption of coal was 4 tons per ton of steel in this period. Only in 1932 was it reduced to 1.58 tons.

During this period, as a result of the increasing demand for pig-iron and steel products, private steel manufacturers appeared on the scene. They had no blast-furnaces; instead they used electric furnaces, and scrap steel and cheap imported pig-iron for steel making. Although the Yawata Ironworks had built up a technological potential, it was not successfully meeting civilian needs. It typified a state-operated mill that gave first priority to government and military needs.

The emergence of electric-furnace ironworks was a response to an increase in industrial demand, and also demonstrated the urgency of the problem of raw materials.

Technological Independence and Dependence on Foreign Raw Materials

The emergence of electric-furnace ironworks corresponded to the fact that Japanese iron manufacturers had a greater capacity for producing steel than for pig-iron. As a result, scrap steel was imported from the United States and pig-iron from India. However, importing from overseas was hindered by foreign exchange fluctuations and problems of unstable supply. Consequently, a solution was sought in establishing iron manufacturing in Japan's colonies and securing raw materials by colonialistic means.

The attempt to transfer technology to the colonies, however, encountered problems because of the low-quality iron ore and the technological difficulties in using these ores. The diplomatic problems that resulted are scarcely in need of mentioning. In this case, consequently, attaining technological independence entailed the need to initiate dependence on foreign suppliers in order to fulfil the first of the five Ms. In other words, the Japanese iron manufacturing industry gained technological independence, but only by virtue of a reliance on overseas supplies. Indeed, this was a paradox of Japan's technology.

Because iron is essential to nearly all industries, the entire network of technologies in Japan intensified its dependence on overseas resources, initially in the 1920s and then in the 1970s as the result of gaining self-reliance in technology at both the minimal and maximal scales.

Japan's independence in iron and steel technology was ushered in in 1920 at the Anzan (An-shan) Ironworks of the South Manchurian Railway Company when a group led by Umene Tsunesaburo (1884–1956) succeeded in discovering a method of pre-treating (i.e. magnetized calcination) low-quality iron-ore. This breakthrough immediately enlarged the range of available resources.

There were two other important technological developments in need of mention. One is the coke furnace developed by Kuroda Taizo (1883–1961). This furnace recovered by-products through a regenerative burning apparatus. It was invented by Kuroda in 1918.

The other is the strong magnetic steel (so-called KS Steel) invented by

Honda Kotaro (1870–1957) through his metallurgical study of alloys. Honda's discovery formed an important basis for Japan's world-leading position in this field.

Parallel to these developments in technology, the Ministry of Finance provided financial assistance to Han Ye Bing, the company that controlled the Da Ye Iron Ore Mine, in 1904, the year in which Yawata began stable operation, to ensure a dependable supply of good quality and low-priced materials.[29]

This financing allowed Yawata Ironworks to import 500,000 to 600,000 tons of iron-ore every year, but it also spelled ruin for Han Ye Bing, China's most important coal and iron supplier (Nakura 1980).

The import of iron-ore and coal from this company once occupied 90 per cent of all imports, but in the 1920s, the supply became unstable, an increased indebtedness led to a loss of autonomy, and the political and social unrest worsened in the Yangtze River areas—all of which adversely affected the company's activities.

During this period, Ishihara Koichiro's Nanyo Kogyo Koshi (later, Ishihara Sangyo) entered the market to take the place of Han Ye Bing. By his own efforts, Ishihara developed the Sri Medam Mine in British Malaya in 1920 and later, he began transporting iron-ore by sea. During his lifetime, he also succeeded in constructing a small combine. In the 1930s, Ishihara was active as a proponent of development of southern Asian resources and of an anti-militarist reform of domestic politics.

In sum, it might be recalled that when Japanese technology arrived at the primary stage of self-reliance, the problem of resources became its Achilles' heel. In an attempt to remedy the problem, technology was transferred to Manchuria and resource development in Korea was undertaken.

One interesting outcome of this policy was that, since, generally speaking, no skilled labour force had formed in these areas, and despite exports of high-level technology, Japan had no greater effect beyond the creation of technological enclaves.

Formation of a Skilled Labour Force

In a study of techology, it is necessary to touch on the problems concerning the education of engineers and the formation of skills. Putting these issues aside for the moment, however, I wish to confine myself here to a brief discussion of the labour force in the steel industry.

The nucleus of skilled workers at Yawata was a group of less than 20 German foremen, who returned to Germany in 1904–1905, and 10 skilled workers dispatched from the Kamaishi Ironworks. The technological isolation peculiar to labour for iron manufacture (the difficulty of converting it to other types of industry) and the heavy physical demands made the transfer of a labour force unfeasible. There was an important difference between the labour force in the Kamaishi area—where heavy snowfall precluded farmers from engaging in agriculture during the winter months, forcing them to take side jobs at the mine, for example—and the labour force recruited in the Kyushu area, where two-crop cultivation prevailed. (The percentage of the Yawata labour force recruited in Kyushu amounted to 80 per cent in 1920, roughly the same proportion as in the beginning.)

At Yawata, former coal-miners, transportation workers (who had worked in transportation on the Onga River until they lost their jobs to the railway), and villagers who had gained machine skills during construction of the iron-works became full-time workers and eventually skilled workers. In the begin-ning (1902), besides the 504 full-time workers, there were many who were indirectly employed on a subcontracting basis.[30] Indeed, the part-time work-ers outnumbered the full-time, a situation that holds true today. The total number of such employees in one year's time reached 600,000 (on the aver-age, 1,500 persons per day).

Characteristically, more workers were assigned to the indirect production divisions than to the direct divisions. Because the major equipment was Ger-man made, importance was placed on maintenance and repair. Second, re-lated machinery and equipment had to be designed and manufactured within the ironworks. This was inevitable since, at the time, the necessary technolo-gy had not been developed.

In the beginning, a majority of workers at Yawata had finished compulsory education (until 1900, four years were required, six years thereafter). Although the relationship between longevity and amount of schooling is not clear, the worker longevity rate at Yawata was low. The authorities of the Yawata Ironworks tried to increase longevity by enlarging lodgings and other welfare facilities, establishing a co-operative purchase organization, and in-troducing a retirement allowance system with progressive rates and a system of commendation for long-time workers.

In general, as operational stability grew, these measures and the salary system were expanded and refined, and the standards to select workers be-came stricter. In 1920, more than 90 per cent of the workers had finished compulsory education. This higher level of educational background in turn contributed to a strong self-awareness and a tendency among workers to view certain management practices critically and to raise demands for improved working conditions. From 5 to 9 February 1920, there was a major strike in which more than 10,000 workers participated. It was interrupted by the arrest of 19 leaders of the strike, but resumed on 29 February and continued until 1 March. The workers succeeded in obtaining a reduction of working hours from a 12-hour, two-shift system to a 9-hour three-shift system. The demand for a "dismissal of incompetent high-ranking managers" and the "promotion of able workers" resulted in an improvement in management and administra-tion.

A system peculiar to the Yawata Ironworks, referred to as the *shukuro* ("veteran") system, was introduced by which highly skilled workers were treated as life-time staff. While by 1930 only seven workers had achieved *shukuro* status, this measure nonetheless helped make a small hole in the hard wall between workers and administrative staff. Worker longevity did not thus, however, greatly improve. During the recession following World War I, the rate of those leaving the mill was stable at 10 per cent. (The national rate for all factory workers was 66 per cent.) The high rate of 36.1 per cent that occurred at Yawata in 1919 was the result of the disappearance of advantageous working conditions and wages along with the economic boom the war had generated.

Iron Manufacturing Technology and Weapons Self-Reliance

As mentioned, the desire for arms independence played a huge part in the development of iron manufacturing in Japan. Before moving on to mining technology, therefore, it will be useful here to consider this aspect briefly.

Japan was compelled to recognize the power of modern weapons when American gunboats forced the country to open its doors. The defeat in the Opium War of China—a country Japan had traditionally looked up to— came as a psychological blow and provoked arguments on coastal defence and a serious study of how to develop the technology for casting big guns and, in turn, for the construction of reverberating furnaces.

A basic problem was the inability of any clan to supply its own fire-bricks to construct reverberating furnaces, let alone the prerequisite blast-furnace; none, moreover, could supply its own iron-ore and coal necessary for operation. There was, clearly, a big gap between the aims of the shogunal regime and the demands of modern technology. To add to the problem, in retaliation against the criticism of the political system lodged by scholars of Western science and technology, the shogunate began a campaign of oppression of intellectuals. Thus, the Tokugawa regime was faced with the contradiction of desperately needing the help of scholars of Western science and yet staunchly opposing the reforms needed to gain and implement their knowledge. Some concessions, however, were unavoidable to adopt a new technology and its related system.

For instance, the Tokugawa shogunate decided to have its own navy; however, because high-ranking samurai, with their retinues of personal servants and retainers and aristocratic habits, could not practically be considered for duty, the government was compelled to select from among only lower-ranking samurai on the basis of ability rather than hereditary status. This rupture of a system based on social ranking that had been in force for hundreds of years in order to effect the transfer of a new, important technology represented a significant risk for the shogunate.

A further concession was the need on the part of feudal lords in hereditary vassalage to the Tokugawas and the shogun's direct retainers to relinquish their monopoly of military technology. They were forced to tolerate attempts by other clans to produce arms and build Western-type warships.

The defeat of the shogunate by the powerful anti-shogunate clans led to a period of ambitious, sometimes blind, technology transfer by the Meiji government. Although, in this period, there existed a national consensus for the transfer of technology for national defence, the burden was not light, neither for the government nor for the country as a whole. Nevertheless, the Meiji government promoted technology transfer based on this national consensus, and its success lent the government a measure of political stability.

To build an arsenal of modern weaponry for national defence required ironworks, for which, in turn, it was necessary to develop sources of iron-ore, coal-mines, and a transportation system. In other words, it was necessary to reverse the order of the "natural course of history" in development, an act fraught, however, with not a few difficulties. Even so, what enabled Japan to effectively reverse the order was the existence of the necessary pre-conditions and a technology that had generally developed to the level of manufacturing technology. Naturally, the pre-existing technology often had to be revised to mesh with the modern technology.

At the time, because Western technology consisted of assembled parts that could easily be disassembled once the principle involved in their construction had been mastered, similar component parts could be manufactured by a full mobilization of traditional technologies.

The technology for iron manufacture was the most complicated and the largest in scale at that time. Although Japan had attained a measure of independence, the difference in level between Japan's technology and the most advanced in existence was still great. Nevertheless, once a technological system—albeit on a small scale and low level—had been established, subsequent technology transfer became easier. The period leading up to self-reliance was marked by difficulties, but the creation of a firm foundation opened the way for future development.

As we noted earlier, underlying the important role attaining "independence in weapons" played in the establishment of a great modern ironworks was the bitter experience of weapons detention in Singapore during the Sino-Japanese War. The standardization of weapons was an important aspect to this effort toward independence. The perception of the need for the mass production of interchangeable parts and a system of standardization of modern weapons dated back to the experiences in the Seinan War (Southwestern Rebellion of 1877, the first civil war in Japan). Both the government forces and the rebels were fighting with imported weapons, but the many kinds of guns and the difficulty of maintaining a regular supply of ammunition created vast confusion.[31]

Consequently, a drive toward standardization in weapons was begun, and it was combined with efforts to establish an ironworks for the purpose of domestic production. Subsequently, domestic production, standardization, and mass production were being institutionalized—principally for small firearms—and, to return to the larger forces at play—in response to domestic and international political developments. These developments had a decisive influence on industrialization as a whole, and the heart of the problem regarding development and technology remains there even today.

Because the technology of iron manufacture was such an important force in Japanese industrialization, I have allocated a disproportionate share of the discussion to its examination. The role of the technologies for textiles, railways, iron-ore, and coal-mines, however, was equally important. Also, shipbuilding technology played an important part in the domestic production of machines for metal and coal-mining and, later, in the production of machinery and machine tools.

8

Transfer of Mining Technology and the Birth
of New Technology

Industrialization and Mining Technology

Modernization in technology had an overwhelming effect on the development of mines. New, Western technology raised their productivity dramatically. In no other area did modern technology have such a profound effect. But it was precisely the combination of traditional and modern technology that gave rise to this result. That is, traditional technology was enhanced by modern technology, and for this reason, this sector of technology warrants a close look.

The Meiji government focused its efforts only on the gold and silver mines, just as the Tokugawa shogunate had done, and for good reason: silver was the basis for Japanese currency, and gold was becoming an international currency.

Under the Japanese silver standard, the price of silver was set higher than its price on the international market. The price of gold was set lower. Western silver, especially Mexican silver, when brought into Japan, had at once a value (purchasing power) three times what it had on the international market. The establishment of an unequal exchange caused an immense amount of gold to flow out of the country. To compensate, the government minted new currency to reflect the true prevailing rate, but the move was opposed by other countries.

Although Japan suffered under this internationally imposed disadvantage, it lacked the political and diplomatic power necessary to resist, and thus was unable to avoid a decline in the value of its currency, deepening inflation and increasing social and economic confusion. Peasant riots throughout Japan were frequent, creating conditions under which the nation risked losing its sovereign power. Unable to resist international pressure, Japan was forced to cope with the imposed unequal exchange and the outflow of gold by industrializing and, particularly, by developing its ore- and coal-mines.

The effort took two forms. The first was the Japanese Mining Law, which the new government promulgated in the fifth year of Meiji (1872). The law precluded foreigners from developing underground resources. Among the government's foreign advisers were some engineers who urged the introduction of highly skilled but cheap Chinese labourers and foreign capital to accelerate development. Due to the great importance the government attached to them, however, it decided their maintenance and active development would best be left to the government alone. More important, Japan still lacked the ability to manage any diplomatic problems that might have arisen had foreign capital and labour been introduced, and the intention was to avoid such a situation in advance.

Thanks thus to government promotion, the mines were developed, and, at one time, Japan was the world's largest copper-producing and -exporting country and the biggest coal-exporting country in Asia.[32]

Pressure from the Western powers on Japan to open its ports in the mid-nineteenth century was aimed partly at securing a supply of coal and water for their vessels. Until the end of the century, the major industrial power source in Japan was water and the major fuels wood and charcoal. Because steam was not a primary industrial power source, Japan had a surplus of coal for export.

The conversion from wood and charcoal to coal (that is, steam-power) in technologies other than the most advanced ones—those of manufacturing, railways, large-scale mining operations, and cotton spinning—was not made as quickly as the subsequent conversion to electricity. Nevertheless, self-sufficiency in energy was a basic condition for industrialization, and it was important in relation to raw materials as a component (M_1) of technological self-reliance.

Compared with the technology used in the newly developed coal-mines, the technology adopted in the metal mines was more complicated and sophisticated. Once coal has undergone dressing, it is then immediately available as a commodity, but ore processing requires a more complicated procedure. Furthermore, since the common practice was to refine the ores at the mine site, the technology for refining had to be added to the technology for mining.

In any case, traditionally, coal had been used only in salt-making areas after the wood supply had been exhausted, and it was transported by boat. Corresponding to the growth of the railways and the greater industrialization following the Sino-Japanese War, however, the newly developing coal-mines experienced dramatic progress. Unlike iron manufacture and the railways, which were developed by the government, the mines became the nucleus for the formation of the financial cliques, or *zaibatsu*.

One such *zaibatsu*, Sumitomo, had acquired refining technology, and was naturally eager to introduce ore- and coal-mining technology. For Mitsubishi, too, a shipping company, involvement in coal-mining was vital, and because coal was an export item, Mitsui, a trading firm, had an interest in Kyushu's coal-mines.

Mining technology and the technology for railways were linked through fuel and transportation needs to form a compound technology. At the same time, out of mining technology grew technological divisions that, from 1920 onward, evolved into separate and independent technologies.

For example, Hitachi (established in 1920), which today has some 8,000 engineers, grew out of the electrical machinery division of Hitachi Mining Co., which engaged in the repair of electric motors. Fujitsu separated in 1935 from the Furukawa Mining Co. (a copper-mining concern) to become a manufacturer of electric wires; it later moved into communications equipment, and is now involved with automation machinery. Although the railways and mining were both characterized as compound technologies, the two developed in vastly different ways.

Related to this, concern has arisen anew regarding the importance of the role played by mining technology in industrialization. In Japan, the government-owned iron manufacturers and railways (private railway companies were established later), together with the private mining and textile industries, were the four leading sectors in industrialization (i.e. technology transfer).

New Technology and Reform of the System

The important contributions modern technology made to mining, besides the technology of refining, were the following.

1. Drainage Technology

The biggest technological barrier to overcome in mining was the disposal of water. Various devices and tools had been invented prior to the arrival of modern technology, but because high-performance equipment breaks down frequently and is complicated to operate, the chief method of disposal was still human labour. Depending on the situation, hand pumps (called "box pipes") and "troughs" were the best means available. The pumps were made of wood and required manual operation, and their drainage capacity was extremely small. Consequently, more than half the mine workers were pump operators, and the older mines, with their deeper working faces, had more of these workers than the newer mines did.

In the Sado Mine, representative of Japanese gold-mines, more than 60 per cent of the workers were engaged in water removal. No worker could bear the labour for more than three years, and, in fact, this hardest of all jobs in the mines was dependent on prison labour. And, for the prisoners, there was no more dreaded labour than this. (In general, until the initial stage of modernization, much unskilled, heavy labour was borne by the nation's prison population.) In 1861, shortly before the Meiji Restoration, of 928 workers at the Besshi Copper Mine (in Shikoku), 447 were engaged in water disposal, and the rest were pitmen. Even so, the water problem forced the abandonment of an otherwise promising pit. Under these conditions, the introduction of a powerful pump could have more than doubled productivity.

Consequently, with the introduction of modern and powerful pumps, a huge labour savings was realized and the water workers were no longer needed, and promising abandoned mines were reopened.

2. Ventilation Problems

As a mine face becomes deeper or more remote from the entrance, the problem of ventilation increases. The excavation of ventilation tunnels as a part of mine development and improvements enlarged the scale of operation, and once machine-powered ventilation was introduced, the mines were freed from the restrictions imposed by natural ventilation. Before these developments, however, the ventilation problem was a serious obstacle to a full realization of the mining potential.

Under the pre-modern contracting system, miners called *yamashi* or *kanako* were paid according to the amount of mineral extracted from the ore and not on the basis of the volume of unrefined ore.[33] This system encouraged the development of a technology for dressing ores at the mine site, but the *yamashi* and *kanako* were not able to develop the technology for an integrated mining system, and, as a result, mines were usually abandoned after the rich veins near the surface had been exploited, leaving huge, albeit less accessible, reserves untouched. This inability to create a comprehensive mining system also meant that only natural ventilation or a similar "burning ventilation" method could be used. However, unlike in the metal mines, the burning ventilation system in coal-mines was extremely limited because of the great danger of gas explosions.

3. Transportation

Transportation within the mines and to the places of demand was also a serious concern. Indeed, since only human and horse transport were available, raw ores underwent primary processing at the mines just because of the restrictions of transportation. The opening of the railway solved this problem.

The possibility of development of transportation within mines from manual labour to horsepower and then to tramcars depended on the physical nature of the mine, the height and width of its shafts, and the design of the route. The full development of a mine with a unified transportation system connecting all the main shafts and its branches would be possible only by implementing modern mining methods. Investments for geological surveys, ranging from the traditional "mountain diagnosis" (the old method of looking over an area and choosing a site according to "feel") to modern boring and feasibility surveys, have to be done in advance, but these were all beyond the scope and financial capacity of the subcontracted mine operators.

Once such modern technologies were adopted, however, they inevitably led to a change in the mining policies inherited from the Tokugawa period.

Returning to the transportation problem, however, one major obstacle related to how the ore was purchased, which determined how the ore was extracted and thus even the shape and condition of the mine itself. The owner of the mines, the shogunate, contracted with the *yamashi* and the *kanako* for their operation and purchased the minerals produced from the ore. Consequently, the operator of a mine would first exploit any outcrops, putting off for as long as possible the excavation of auxiliary tunnels necessary for water drainage and smoke removal. In search of only the rich veins, operators tunnelled tortuous courses that varied in height from section to section, thus creating problems for transportation and safety maintenance. As a consequence, mines everywhere developed in the shape of a honeycomb. In 1751, the Ashio Copper Mine had as many as 1,484 entrances. The reasons for the decline of the mines toward the end of the Tokugawa shogunate are thus easy to determine. This very inefficient operational approach meant that even rich mines were abandoned after an average of only three years in operation (Murakushi 1979).

The establishment of an integrated system of mining, transportation, dressing, and metallurgy thus began to conflict sharply with the system of purchasing ore. To achieve full rationalization, it was necessary for mine owners to integrate the various workers and jobs into a single, well-organized, directly controlled production process. This became possible only at several nongovernmental mines after the Meiji Restoration.

However, modern technology was introduced for the first time at government-owned mines, to which as many as 87 engineers, pitmen, and machine operators were invited from Great Britain, the United States, France, and Germany. This marked the start of modern integrated mining operations in Japan.

Among the new procedures introduced at this time, the most technologically noteworthy was step mining, which was introduced in 1868 at the Sado Gold Mine and the Ikuno Silver Mine. Six years after its introduction, the method had spread to all the leading mines in Japan. Under this method, the shafts were equipped with a hoisting whim and a rail drift, and new shafts were systematically connected and opened as necessary.

This method enhanced tremendously transportation within shafts. Even though transportation by tramcar remained dependent on manual labour and horsepower, efficiency was strikingly raised. By virtue of this new system, ore was transported to the outside by the hoisting whim—a far cry from former times, when miners were forced to crawl on hands and knees and considered themselves lucky if they could walk in a half-sitting posture. With the introduction of dynamite in 1878, efficiency in opening shafts and drifts improved dramatically. From these developments, it becomes clear how essential structural reforms are to technological transfer.

4. Pre-Conditions

The rapid gains in productivity realized by the introduction of M_2 components for drainage, ventilation, and transportation were made possible by the pre-existence of M_3 and the other five Ms. The existence of the other components provided important support for the introduction of M_2. Thus, technology transfer, the introduction only of M_2, is inexpensive and effective when fully supported by the four other components. At the same time, M_2 essentially changes the principal composition and role of each of the five Ms.

Regarding the manpower component (M_3), whatever changes might be introduced by M_2, it is important to insure that the quality of labour is protected, that is, the high level of traditional skills and the organizations of workers that help to maintain this level. Indeed, in the case of the mines, worker organizations constituted an important pre-condition of M_3.

The mines had two kinds of workers. The first was skilled workers called *jikofu* (indigenous, or local, mine workers), who had been trained from an early age for work in the mines. While most of the workers in this group were not exclusively miners, they had long experience and profound knowledge of the mines in which they toiled.

Mine work was often performed as an extension of non-mine work by small, family groups made up, for example, of parents and their children, husbands and wives, etc. Women and children were usually engaged in such auxiliary work as carrying out the ore. Because the work was dangerous, a sense of absolute trust and mutual reliance was essential to maintaining efficiency and safety.

The second group consisted of skilled workers called *watari*, or "floating" pitmen, who, as the monicker suggests, did not settle at any one mine. They moved from mine to mine at intervals of several months or a few years, enriching their experiences and accumulating more skills and also diffusing skills along the way.

Many had special skills, for example in water removal, smoke dispersal, or rock boring. Such skills were not necessarily required at all times at any one mine, but they were vital when a mine was being opened or during a period of recovery from an accident. Because of their specialized work, they were the "rich" of the poor, who earned high wages, had high mobility, and usually were unmarried. Their strong orientation toward special skills endowed them with a different way of life from the local mine workers described above.

A nation-wide organization of mine workers, called *tomoko*, acted as an information network for those in need of the specialized skills of the *watari*.[34] Before admission into the *tomoko* union, an unskilled worker would be engaged in auxiliary tasks (such as the transport of ore and debris) to become accustomed to the work in the shafts and drifts. Under the direction of skilled pitmen (called *sakiyama*), the unskilled worker accumulated skills as a pusher (called *atoyama*), covering the whole process of production, including excavation, simple dressing, and accident prevention. Once an *atoyama* had acquired a specified level of skill and experience, he became a licenced pit-man and was authorized by the *tomoko* to be a skilled pitman, thus also qualifying as a master, or foreman.

As a *sakiyama*, he assumed a quasi-parental (superior-subordinate) relationship with pushers and others to form a unit for production activity and to subcontract a pit face. Generally, the *sakiyama* and his men were under the control of a principal contractor, called the *hamba* (or *naya*), for entire veins. When integrated mine operations were developed through modern technology, the *sakiyama* became directly contracted workers.

The Japanese mining industry developed rapidly at the end of the nineteenth century. As a leading industry, it experienced a sharp increase in the number of pitmen. By 1894, the total employment population reached 4,570,000, of which 100,000 were mine workers (metal-mine workers totaled 55,000 and coal-miners 43,000). The number of mine workers peaked in 1917 (176,000 at the metal mines alone). In 1922, however, this total declined drastically, to only 40,000, because of mine closings caused by the economic recession after World War I and the development of rationalization.

In the process of modernization, the pitmen's tools, the chisels and hammers of the era of the "gold-mining carpenter," became obsolete, and it now became essential for pitmen to master such modern technology as drilling with jumpers and blasting. At the same time, the acquisition of multiple skills was replaced by the need for a single skill, which led to a division of labour among pit workers, and, as a result, the *tomoko* union changed to reflect this new specialization in skills, and is said to have "begun to assume a distinct character."[35] This change represented the second phase in the transformation of mining technology.

Another important function of the *tomoko* union was for mutual assistance. Although the change in mine work caused a change in the nature of contracting, the union's policy of social security and welfare, to which nothing similar had been established in Japanese society as a whole, was maintained for the protection of its members. In addition to the assistance provided for wedding ceremonies and funerals, assistance and relief were extended to pitmen unemployed because of a mine closure in the form of providing information on employment opportunities, and aid was given to persons suffering from occupational diseases or injuries. These functions later formed a contact point with the labour union movement. In not a few cases, the *tomoko* and labour unions, by providing mutual support and exchanging knowledge and expertise, ensured the existence of both, particularly when labour movements were being oppressed under military rule. Also, the influence of the *tomoko* was widely acknowledged during the period of recovery of labour unions after World War II.

In general, because the *tomoko* were made up mainly of pitmen, whose skills found favour with the development of step-mining technology, the *tomoko* prospered. However, the indirect system of employment through subcontracting changed to a system of direct employment. The change in mining technology caused by the introduction of small rock drills meant the loss of the *tomoko*'s technological leadership, and the on-the-job training and schooling programmes given by individual companies began to assume the leading position in skill formation. Also instrumental was the transition from the selective mining of rich ores to the mass mining of ores that came with the introduction of the mechanization of dressing (and especially the introduction of the flotation method in the metal mines).

The technological changes in mining paralleled the transformation of the labour and management system from the system of contracting for the production site to a system of directly employed workers. Indeed, they corresponded to the complete transplanting of the system of modern mining technology. As a result of this process, the position of pitmen, the number of which had increased as many as 10 times in the 30 years from the end of the nineteenth century, decreased relative to the entire labour force in the metal and coal-mines. The number of factory-type workers grew to outnumber the pit workers, and their roles in the operation of the mines, which were now using a compound technology, became indispensable.

The age of the *tomoko* had started gallantly when modern technology was introduced and declined with the development of modern compound mining technology. However, the role of the *tomoko* as a pre-condition for M_3 was great in that the systematic development of modern mining technology was supported by this autonomous organization devoted to workers and their skills. Modernization of the mines thus depended on the *tomoko* as a pre-condition; the next step, independence of mining technology, required the domestic production of machines that were being imported.

From Importation to Domestic Production of Mining Machinery: Independence in the Related Sectors of Technology

Introduction of Technology—The Case of Ashio

Compared with coal-mining technology, the technology for metal mining poses greater challenges because of the complicated refining process and related equipment that is required.[36] The process of the establishment of a modern technology for an integrated production system from mining to refining was in fact a process of gradually amassing new technology. In technological terms, it was the formation of an enclave at the production site.

Mechanization and a change in the source of power started in water removal, ventilation, and transport. These improvements increased production and allowed for operations of larger scale and at higher speeds. But the pit work still required manual labour. An increase of production could thus be realized only by means of an increase in pitmen. It was typical that the mechanization of excavation was the most delayed in Japan.

The Ashio Copper Mine, which had been expropriated by the Meiji government after the Restoration, fell into private hands soon after; the government regarded it as being too far gone for reactivation. Although it had once produced an average of 1,100 tons per annum for more than 70 years (from the end of the seventeenth century), it was nearly completely abandoned as a result of extensive cave-ins, which could be attributed to the random excavation of only the richest veins that was typical under the afore-mentioned technological and structural limitations prevailing under shogunate management. In 1876, the mine was actually, by default, in the hands of the subcontractors, because the original mine owners were no longer able to meet the wage payments. When Furukawa Ichibei (1832–1903) bought it in 1877, it was like a vast honeycomb, and Furukawa was rumoured to have gone mad.[37]

But it was Furukawa who revived the mine to make it one of the most productive copper mines in Asia. He had international business experience as an executive member of a silk-trading firm, but he learned the mining business at the production site. His success in managing the Ani and Innai mines, which had also been sold to him, clearly revealed his excellent managerial skills.

His talents as an outstanding manager were shown not only by his recognition of the true value of the old Ashio mine but also—at the time he acquired the mine—by his hiring of gifted young university-trained engineers, extremely valuable human resources at that time, and his decision to buy imported machines. In other words, the establishment at a stroke of M_1 through M_4 had a tremendous impact on the company's development.

What remained for the company to attain was M_5 (markets). The beginning of the 1880s marked a world-wide overproduction of copper because of the development of copper mines in several western US states. But the substitution of iron telegraphic wire with wire made of hard copper in Europe in 1886 created new demand for copper and transformed its market. Encouraged by this new trend, Jardine, Matheson and Co., a concern well-known from its wide-ranging activities in Asia, proposed a long-term contract with Furukawa in 1888, by which Furukawa acquired, in a single stroke, a full set of M_5, which, further, made possible a complete renewal of the mine's facilities.

Furukawa succeeded in roughly 20 years' time in modernizing the Ashio Copper Mine. The first step was a careful survey of pit faces and the adoption of a direct excavation method aiming at rich veins. This method, designed to open mines from the gangway for drainage and hauling, was similar to the adit-opening method used in the West and was aimed at topping rich veins by speeding digging from the lowest level (Murakami and Hara 1982). Furukawa had successfully adopted this method at the Kusakura Mine. Five years after the purchase of the Ashio mine, the miners discovered—using this method and after digging 140 meters—the first rich vein; two years later, a second one was uncovered.

At the beginning of Meiji, every government-operated mine was unsuccessful because emphasis was given to equipment and mechanization at the expense of the development of new sources of ore, the life of a mine (Murakami and Hara 1982).

In contrast, Furukawa adopted a comprehensive development programme based on the modern method of opening a gangway; nevertheless, the company otherwise remained dependent on traditional technology. In comparison with Sumitomo, a pioneer in copper mining, Furukawa was anxious to introduce modern technology, but realization came only in later years. Even as late as 1881, Furukawa was still relying on manual crushers and leather foot-blowers instead of steam-powered crushers and ventilators (Hoshino 1982).

On the other hand, Furukawa did use blasting powder to excavate pits and introduced rail cars for transporting earth out of the mine. In all other aspects, the old methods were still in place: manual labour was the main force for digging and for hauling.

In 1885, Furukawa started to excavate the main adit, which took, indeed, 11 years to open, but, once completed, it ensured the life of the mine for roughly 100 years. The year 1885 was important in other ways for the Ashio Copper Mine: the rock-drill was introduced and the boiler-type pump was added to the manual pump for drainage. For the large-scale excavation of the main adit, engineers and equipment from several Furukawa-owned mines were used (Shoji 1982).

At the end of the same year, however, sudden flooding destroyed the mine's drainage equipment. Although the mine yielded a record 4,090 tons of copper, there was not enough capital to make possible the introduction of large-scale modern equipment. The 1888 contract with Jardine, Matheson, however, brought the company the fantastic amount of ¥60 million, which enabled it to buy modern equipment and eventually report a sharp rise in output.

The first item of equipment introduced was the steam drainage pump. Before its adoption, 10 hand pumps of two different types had been used to pump the water up gradually from the deepest levels of the mine to the shallower pits, where it would be concentrated in a single pit before finally being pumped out of the mine. More than 310 workers and 24 hours were required for this operation. But the steam pump completely eliminated the need for this. By 1890, a hydraulic power system was set up to operate an 80-hp pump for drainage, a 25-hp hoist, and a 6-hp generator for lighting. This completed electrification within the pits, and transport and drainage at the shaft were also electrified.

The refining process was also introduced in 1890. The four metallurgical processes consisted of calcination, melting, kneading, and refining; a soft-charcoal furnace was used for the calcination (15–20 days) and for deriving crude copper by the oxygen-burning method; and a round-bottomed furnace for melting ores (called "Yamashita blast," 6 hours). The whole process required 32 days and produced one ton per furnace. Not only was the process slow but the furnaces had to be repaired after every batch. However, with the adoption of the reverberating furnace for calcination, the square-type water-coated furnace for melting, and the revolving furnace for kneading (the principle was the same as for the *mabuki* refining process), the procedure was reduced to only two days and allowed for continuous processing in large quantities.

In 1898, electrolytic refining commenced with the construction of 116 electrolytic cells. Production of copper wire had begun the previous year in Honjo, Tokyo. Thus, a system of continuous production, from mining to an integrated metallurgical system, was established in 20 years' time.

In addition to the introduction of tramcars for transport and electrification in the mines, electric railways were laid between the major mine entrances and the refineries. Outside the mine, though, delivery via the Joshu and Nikko routes depended on pack-horses until 1888, when a horse-drawn carriage was introduced on the Nikko route (the width of the road was about four metres). In 1890, an overhead ropeway system using steam-power was installed, and in 1892, the horse-drawn tramcar was introduced on the Joshu route, and in 1893 on the Nikko route.

By 1890, the facilities at the Ashio Copper Mine had been modernized; the mechanized system in the pit had been electrified and the fuel for refining had been changed from charcoal to coke (completed in 1893). Despite these developments, however, as just mentioned, the transportation network outside the mine was undeveloped, and under such conditions it constituted a good example of an enclave of amassed technology.

With the opening of the national railway's Ashio line (for steam locomotives) in 1914, however, this technological enclave was "thrown open" and the technological spin-offs from the mining industry, which constituted a compound technology, that had been prepared could now flourish. The year 1914 was also the year in which drills were first produced domestically, at the Ashio mine.

A quarter-century had passed since modern mining technology had been introduced at Ashio. The mine was blessed with favourable conditions that allowed a rapid transition from technological transfer to technological offshoots.

Nevertheless, the long time required before technological self-reliance was attained (even now not fully realized) is noteworthy. Technological self-reliance was influenced by the circumstances that surrounded copper mining. Fluctuations in the international prices of copper, a primary product, particularly affected the Ashio mine, causing it to swing between boom and recession. Domestically manufactured machines (and the company's own home-made machines) were introduced at Ashio to help alleviate the difficulties caused by these fluctuations, but this had to wait until the machine industry in Japan had achieved a certain degree of progress. On the other hand, the laying of tracks for electric cars within the pit preceded the electrification of the national rail routes and the founding of the Yawata Ironworks. Finally, the discovery at Ashio of a huge deposit of high-quality Kajika ore greatly aided in coping with the depressed copper market.

The formation of a technological enclave is fatal to mine development in late-comer countries; in Japan, it contributed to severe copper poisoning and environmental destruction. The sulfurous acid gas emitted during refining is itself poisonous, and combined with rain, it forms sulfuric acid, which is highly damaging to forests. At Ashio, besides raising the river-bed and increasing flood damage, the dumping of slag from the mine made uninhabitable the mountain villages downwind from the refinery and destroyed rice fields of the villages downstream from the mine.

Environmental and pollution problems were the negative by-products of industrialization, problems that resulted especially from the technological enclave.

Domestic Production of Mining Machinery

Early on there was a strong tendency to rely, as far as possible, on domestically manufactured machinery. Thus, as early as 1877 the government-operated Miike Mine was purchasing finished machinery from the government-run machine tool plants (located at Honoura, Nagasaki; Kobe; Akabane; and Kamaishi). Also, the importation of foreign drilling machines was resisted in favour of purchasing from the government plants. Nevertheless, at this stage complete dependence on domestically manufactured machines was impossible, and so of necessity, Miike had also to rely on imports.

Miike established a foundry in 1882 and began repairing and producing pit machines and machinery for ships connected with the company. By 1887, the company had five factories: a smithery, a cannery, and plants for finishing, boiler making, and fabricating wooden models. It also made experimental steam-powered pumps and hoists.

According to Kasuga (1982), the foundry progressed into the later Miike Works through three stages: (1) 1888, when the government sold the mine to Mitsui, to 1899; (2) 1900 to 1909; and (3) 1910 to 1918.

The activities characteristic of the first stage were the repair of machines and the production of wooden transport vessels. The company employed 13 skilled workers and others from the Mitsubishi dockyard in Nagasaki, under whom—to secure and train a labour force—paid and unpaid apprentices were assigned.

At the beginning of the second stage, the company had in its employ British engineers, who designed the endless rope haulage system used for transportation in the pit. While learning design methods, the foundry workers made the leap from mere repair or improvement when they manufactured the davy pump at about the time a new mine was opened. Production costs, including the construction of the workshop, were 25 per cent lower than when the British pumps had been used. Imported goods cost three times as much as domestic products because of low yen values in foreign exchange markets.

In 1893, a three-ton iron-melting furnace was imported to be used to manufacture 30-inch special duplex pumps. They were inefficient, however, and the company imported Washington pumps and attempted, through imitation, to manufacture them themselves. By repeating the process of importation → imitation and trial production → acquisition of technology, the company acquired the ability to produce steam-pumps and transporters, but it did not yet have the technology for the design or the mass production of machinery. By the end of stage two, however, it was accepting orders from coal-mines in Kyushu to produce small- and medium-sized equipment to be used for drainage and transportation. The company thus was able to expand to producing not only for itself but also for other companies, and it had progressed from imitating design to creating its own to meet local conditions and needs.

The third stage corresponded to full mechanization in all the coal-mines and the advance of electrification. Those who had been engaged solely in electrical equipment repair and improvement were now producing electric hoists (1904) and electric endless rope haulage systems (1906). The incentive for these changes had come from the high prices of imported products, which were 40 to 80 per cent higher than the cost of producing them in-house.

In 1912, the greatest difficulties in producing cylinders were at last overcome, and a gas engine was successfully produced. This instilled tremendous self-confidence in the company's manufacturing abilities. It also promoted the electrification of the Miike Coal Mine and facilitated enlargement of the coke plant and the establishment of dye-making and zinc factories. It also marked an important starting point for the entry of technology in other, related sectors, especially in the area of machine manufacturing.

A similar process occurred at the Ashio Mine. Furukawa Electric Company was founded in 1896 to manufacture electric wires. In 1923, the company and Siemens, of Germany, founded Fuji Electric Manufacturing (later Fuji Electric) to embark on the production of heavy electrical machinery and then communications equipment. Similarly, Sumitomo's Besshi Copper Mine established Sumitomo Shindo (later Sumitomo Metals) in 1898 to process copper; Sumitomo Denko was set up in 1911 to manufacture copper wires, and Nippon Electric was established jointly with Western Electric of the United States in 1899.

The division for the repair of electrical equipment at the Hitachi Mine manufactured small electric motors. Later, it developed into Hitachi Limited, a heavy electrical equipment manufacturer. Hitachi subsequently moved into light electrical appliances. The Furukawa-affiliated Fuji Tsushinki and the Sumitomo-affiliated Nippon Electric Corporation became the two biggest manufacturers supplying telephone-related equipment to the Ministry of Communications (later Nippon Telegraph and Telephone Public Corporation). These three companies today comprise the leading electronics manufacturers. Significantly, all three began as repair divisions of mining companies. My theory of the five stages, from technology transfer to independence, is based mainly on an analysis of this sector.

9

Traffic and Transport Technology—Road, Railway, and Water-borne Transportation

Modernization and the Railway

The government-operated Yawata Ironworks produced iron and steel for both munitions and for railway- and bridge-construction materials.

The mission from Russia that visited Nagasaki in 1853 to request the opening of the port brought with it a model of a locomotive that used alcohol for fuel to officials of the shogunate and, to everyone's surprise, claimed the full-scale locomotive could pull several carriages and carry 500 persons up to 280 *ri* (1,081 kilometres).

In the following year, the US Navy fleet visited Japan for the second time and brought with it a large model of a locomotive, along with telegraph instruments. The American locomotive was purportedly 2.4 metres long, 1.5 metres wide, and 3 metres high (see Harada 1979, 1980), and it pulled a car on a 109-metre loop railway with a 550-millimetre gauge. The story is told that people could get on the roof of the carriage. In any event, there was great excitement and curiosity among the onlookers. In the same year, the Japanese translation of P. van der Berg's *Eerste Grondbeginselen der Natuurkunde* (1884) was published, in which the mechanism of the locomotive was systematically described.

Thus, considerable knowledge of railways had accumulated by the end of the Tokugawa period, and it is said the shogunate had plans for a railway. It was not until the Meiji period, however, that the plans were realized: A man named L. C. Portman of the US legation in Japan pressed the new government for confirmation of the concession to lay a railway between Edo (Tokyo) and Yokohama that he claimed he had obtained in 1867 from an official of the shogunate's government. The new government, however, had no policy concerning a railway and was startled by the request.

In a hasty decision, the government declined to confirm the concession. Starting with the laying of rails between Tokyo and Kyoto, the government intended to construct and to operate all railways in Japan independently.

This decision brought on a diplomatic dispute with Portman and the US legation. The new government, while over-sensitive to diplomatic problems generally, was able to refuse because it had the strong support of the British minister to Japan, H. S. Parkes. The government's policy to build and operate a railway independently suited Britain's desire to eliminate its competitors, and, through Parkes's mediation, an agreement was made to raise a loan for construction on the London market and to introduce the technology from Britain. Applications were accepted for the concession to lay a railway not only between Tokyo and Yokohama but also between Osaka and Kyoto.

Japan's decision to inaugurate a railway system was taken in direct response to foreign pressure, in particular, to British diplomatic and commercial interests, and this was opposed by such technocrats as Inoue Masaru (1843–1910), who advocated Japanese "government construction and government operation."[38]

Subsequent to the Meiji Restoration, Edo was renamed Tokyo and made the new capital, but Japan actually seemed to be a country with two capitals: one in Tokyo and the other in Kyoto. Consequently, the construction of a railway connecting the two cities was necessary for political stability.

Thus it came about that, although railways in Europe had developed as industrial railways on the basis of economic development after the Industrial Revolution, the railway in Meiji Japan began as a political railway and the industrial aspects arose later.

Ninety years after the first Japanese train puffed its way between Tokyo and Yokohama, the Shinkansen (the so-called bullet train on the New Tokaido line) began operation concurrently with the Tokyo Olympic Games in 1964, and Japanese railway technology thus became among the most advanced in the world.[39] The Shinkansen was an inspiration to people in the world's railway industries, people who had been discouraged by the industry's decline under the influence of automobiles and aeroplanes. As a consequence, Japanese railway technology has come to share a part in the export of technologies.

The role of the railway in Japan's modernization is so great that it may be called the "keystone of modernization" (Harada 1979, 1980). Besides its effect on economic development, the railway—along with schools and the military—is said to have been a principal contributor to the Japanese habit of punctuality. In fact, the railway helped in many ways to spread modern Western civilization throughout Japan.

Iron manufacturing and the railway each constituted a cornerstone of Japan's technological development. Conversely, the railway also constituted a prerequisite for colonization, and in the Japanese colonies, the railway was an important tool in maintaining authority.

The Transportation Network

The railway symbolized the "civilization and enlightenment" policy of the Meiji government. It was the symbol of a new, modern government for Japan and its people. Indeed, the railway revolutionized transportation, but it also accelerated the confusion predominant in those days among the diverse means of transportation. Japan had not experienced a full-scale development of the horse carriage when the railway was introduced (Yamamoto 1979, 1980). Before the Meiji Restoration, vehicle transportation on the highways was prohibited. Travel was allowed only by foot, horseback, or palanquin. Heavy cargo was usually transported by boat on rivers and canals. For moving goods, people depended on the coastal and inland waterways as main transportation routes; water-borne transport was quicker and cheaper and had a greater capacity than road transport. Consequently, roads were secondary.

Nevertheless, the roads were well-developed; all villages were connected by roads, though they could not be used in all kinds of weather. With the roads and the well-organized network of boat transportation, Japan experienced high traffic density in the nineteenth century.

Even by the 1980s, statistics reveal, total road mileage in Japan had increased but 20 per cent since the Meiji Restoration. The old roads, however, had been constructed for travel by foot, horseback, and palanquin; thus, their widths and grades were unsuitable for vehicles; stoned-paved roads were rare. Although the increase in total kilometres has been small, the total area of roads has expanded several dozen times. Motorway construction got underway in earnest in the 1960s, and before long a network of highways blanketed the country.

After World War II, the Occupation forces were surprised by the poor roads in Japan, but this was due to Japan's short history of automobile use. There had been no full-scale development of carriage transportation to serve as a preliminary stage for the age of motor vehicles. Because the diffusion of railways was so quick, Japan was an advanced country for railway travel but one underdeveloped for road travel.

The highways (there were five) that had been directly administered by the Tokugawa government were well maintained, and many Europeans left records of favourable impressions. However, for security reasons, the Tokugawa government limited travel and prohibited the construction of bridges.

The new Meiji government, however, granted freedom of travel and transportation, allowed the construction of bridges, and promoted the development of roads. It even allowed toll collection for a limited period to cover construction costs for the bridges, which were built and financed by the private sector. In one sense, the Meiji period had introduced the first road age in Japan nearly a century before the second one. Road and railway traffic were added to water-borne traffic, and a period of mixed transportation thus commenced (Yamamoto 1979, 1980).

Regarding road traffic, the palanquin disappeared; in its place, the carriage appeared, and a large cart (called *daihachi-guruma*) and the horse. The horse-tramway was newly introduced, and travel by horseback was still popular. The rickshaw, a Japanese invention, also appeared. In a period of approximately 10 years, a great variety of transport means had emerged. The "express" route between Tokyo and Kyoto (495 kilometres) consisted in 1878 of 1 section of railway, 7 of horse-drawn carriage, 1 of rickshaw, and 4 of walking, creating a very patchy traffic route totalling 13 sections that required 60 hours of travel.

In 1889, the railway between Tokyo and Kobe was opened to through traffic, and extensions of trunk railway lines spread the network throughout Japan and thus alleviated much of the patchiness of the system. Nevertheless, because Japan is an island country, patches have remained in marine traffic where ferry routes are operated by the railway companies. The Kammon undersea tunnel was constructed in 1942 to connect the main island of Honshu with Kyushu; its extension on the other end of Honshu, connecting that island with Hokkaido, was completed in 1988.

Digressing momentarily, mention might be made of marine transport, which was an important element of industrialization and which developed in relation to railway development in Japan. Japan's closed-door policy had included a prohibition against the construction of ocean-going ships, and, consequently, Japan had no technology for constructing and operating large vessels. Thus, one of the first steps taken by the Tokugawa shogunate in establishing its defence policy was to purchase from the Netherlands a steam-powered warship (300 tons, 3 masts, 100 hp) and training for its crew.

The opening of Japan's ports allowed foreign vessels to enter into the coastal transport business. Japanese ships could not compete with these vessels in speed and carrying capacity, and consequently in freight rates. To cope with this situation, a semi-governmental shipping company was organized, using government-owned ships (which were foreign made and which the Meiji government had inherited). The business went bankrupt within a few years, however, and in 1875, the government adjusted its policy to one of fostering a powerful private steamship company. In this way, Mitsubishi acquired 13 government-owned vessels and a subsidy of Y2.53 million for 10 years, enabling it to overtake its competitors, the Pacific Mail Steamship Co. (US) and the Peninsular and Oriental Steamship Co. (Britain), in domestic coastal transport, and, in 1876, to enter into service between Yokohama and Shanghai. From that year, Mitsubishi was obliged to establish a merchant marine school and was granted a 14-year subsidy for its shipping operation.

A long time was necessary for shipbuilding technology to become independent, and it is noteworthy that its arrival at self-reliance precisely coincided with the development of self-reliance in railway technology.

Although Japan had the ability to create basic design, it did not have expertise in the details of design and the processes or specifications of shipbuilding. It therefore placed orders for ships with foreign countries and, in an attempt to master the technology, requested that delivery include a complete set of drawings and other pertinent documents.

Returning to the railways, railway technology was of great importance because its spread led eventually to the establishment of machine manufacture throughout Japan. Machine manufacturing technology was transmitted through the individual manufacturing divisions that developed in the railway industry, and many local plants became the site for the accumulation of technology.

For example, the Hamamatsu plant became the birthplace of the No. 1 model of the Type C51 locomotive, used on the special express train Tsubame (swallow), which marked the golden age of the Japanese National Railway. This was in 1919, seven years after the wholly domestic production of locomotives had been achieved. Although electrification had been completed between Tokyo and Numazu, a steam locomotive was used for this "star" train because of its high reliability.

It is significant that a locomotive of this stature was manufactured in a regional plant. In 1920, Hamamatsu had a population of only 65,000, but on the eve of World War II it had become a major centre of the machine industry and of radio-wave technology. After the war, makers of famous musical instruments and motorcycle manufacturers sprang up in this area, the seeds for this industrial development dating back to the establishment of the railway's machine manufacturing division. Further in the background was the technology of woodworking and the manufacture of weaving machines, but our interest lies more with an examination of the machine manufacture division.

The railway represents a compound technology; it embraces complicated, precise mechanisms of several branches of technology, such as construction, communications facilities, signalling, electricity, supplying of coal and water, and the utilization and administration of machines, tools, and power. It symbolizes well the formation of a national network of technology because the scale of its technological linkage is extensive and its level is high. At least this has been so in Japan, and for this reason, the focus on self-reliance in railway technology is perhaps well justified.

Issues in Railway Policy

Railway technology consists roughly of the following five systems: (1) operation (running, communications, and signalling); (2) maintenance (track maintenance, repair); (3) production (design, production); (4) construction (surveying, laying track, including construction of tunnels, bridges, and facilities for transmission and distribution of electricity); (5) station services (for passengers and freight, planning and administration of running and marshalling, safety).

Because of the wide range of these systems, one problem is to interrelate the technological independence attained in one sector with that in others. It is thus apparent that technology can be developed only in a spiral through the convergence of linkages.

Railway technology was imported for its usability—not so that it could be reproduced domestically—and its transfer depended on the pre-existence of a native civil engineering technology. Foreign engineers were in charge of basic design, and according to their records, the Japanese level of surveying was already high and the Japanese were quick to master new technology. This was in thanks to the good use made of the stock of traditional technology, the essence of which lay in castle construction.

The standard gauge (1,435 millimetres) used in European countries was not adopted in Japan. Rather, the narrow gauge, of 1,067 millimetres and called the "colonial type," was adopted. It was in use in India (between Bombay and Turner), Australia (Flinderstreet and Melbourne), and in China (Shanghai and Wusung).

It is not certain why the narrow guage was adopted. One reason may be that the politicians simply did not know railway technology, or it may have been, since the money and materials were procured in England, the Japanese authorities were forced to accept the standards of railway construction adopted in Asia by the English. The cost of the narrow gauge, less than that of the wide gauge, was perhpas an important consideration at the time.

However, as the value of the railway as an industrial artery increased, a call arose for revising the gauge so that the efficiency and high speed possible with the wider gauge could be realized. In general, railway engineers and bureaucrats advocated the wider gauge. As with the Yawata Ironworks, the formal English name of Imperial Government Railway expressed Japan's desire, as a late comer, to catch up with the forerunners and take its place among them.

Because a change of gauge included the improvement of tunnels and bridges, it was advantageous to start the conversion as early as possible. Changing the gauge after completion of the network would be impossible because of the greater difficulties and financial burdens this would generate.[40]

On the other hand, there were those concerned with the role of the railway in local development who rejected this technology-first attitude. They maintained concern ought to be over an extension of absolute mileage rather than technological quality. Railway policy thus became a political issue: change over to the international standard with an eye to technological independence or give priority to development, in other words, vertical or horizontal expansion of technology.

In the 1910s, in parallel with the development of party politics, the matter of the railway became a serious election campaign issue—that is, a rallying point for mobilizing local interests—among the political parties. One result was the construction of the "hooded crane" of the Ofunato Line (Iwate Prefecture), a U-shaped line that was three times longer than the shortest possible route promoted for the interests of their constituents by powerful Diet members. Similar political decisions were repeated after World War II, which constituted one reason for the overall decline of the Japanese National Railway.

Regarding railway technology, the role played by the military at the initial stage of the opening of the railways should not be overlooked. When the decision was made to establish the railway, it was complicated by diplomatic issues, and the new government's financial basis, burdened by the debts inherited from the former government, was weak. The Emperor, although the sovereign, did not have the imperial military forces under his direct control and was guarded by soldiers provided by the powerful anti-Tokugawa clans.

Under such circumstances, a few leading figures of the Meiji Restoration, such as Saigo Takamori (1827–1877), insisted that the railways were the "way to financial ruin" and that it was more essential to organize the nation's armed forces.

The military authorities recognized the importance of the railways, however, after the Southwestern Rebellion (1877), but they opposed a plan to lay a trunk line between Tokyo and Kyoto along the old established Pacific coastal route (called the Tokaido) because of its vulnerability from the viewpoint of defence. They insisted instead on a plan for a line through the central mountains (the Nakasendo route), which would have required a huge outlay of money. In the end the military conceded for technological and financial reasons. And from 1887, it fervently insisted on the adoption of the wide gauge. But in 1898, this insistence suddenly stopped, and the military proposed the adoption of standard specifications in design, operation, signals, and administration and had these specifications effected.

The year 1889 was an important year for the Japanese railway system: it was the year traffic opened between Tokyo and Kobe. Also at this time railway construction and operation by the government reached a deadlock for financial reasons, and when private railway companies sprang up, they won the concessions to lay Japan's longest line (from Ueno to Aomori) and the lines beyond Kobe.

In 1890, the total mileage of private lines was 1,364 kilometres, as opposed to the government's 885 kilometres. This mix of government and private operations brought on the need for adjustments in railway construction policy—from both military and commercial standpoints. It was essential and inevitable that all railways in Japan be standardized under a single set of technological specifications (in sharp contrast to the situation in some other Asian countries). In this the interests of the railway technocrats coincided with those of the military technocrats. Thus, technological compatibility and consistency became the integrating feature of the above-mentioned five-system compound technology. Before this standardization, all railway companies had been obliged to allow the government free use of their trains in times of emergency or disturbance.

In 1890, Japan was struck for the first time by a capitalist economic panic. Some private railway companies went bankrupt as a result, and there arose calls from the private sector for nationalization. The leadership in railway construction thus fell again into the hands of bureaucrats (military and railway), hence advancing the cause of nationalization.

Railway construction and operation by the government thus developed, leading eventually to the formation of a Japanese national railway system. However, the national railway system was confined to the trunk lines, and the private railways survived.

After World War II, private railway companies prospered and the national railway declined. One exception to this was the Shinkansen, whose operation co-efficient is 60 per cent (that is, the cost as a percentage of fare is 60 per cent). Many local lines have suffered deficits, and the Japan National Railway carried a gigantic deficit over many years, eventually forcing a fundamental reform.

Original Design and Production by Imitation— The Road to Self-Reliance

The construction of the line between Kyoto and Ohtsu, which was begun in 1878, was difficult because of the hilly terrain, rivers, and other obstacles. Inoue Masaru, mentioned previously, assigned foreigners as advisers, bridge and tunnel designers, and used only Japanese for the actual laying and construction. As a result, the line was built exclusively by Japanese. For the construction of the tunnels, for example, Japanese mining technology (the skills of the miners at the Ikuno Silver Mine) was made use of.

Here, then, is an example of a rational combination of new design technology from abroad and the stock of traditional technology, which proved to be wise management in the process of transfer.

The army's advocacy for standardization of the railway system was a policy of the unification of technology. Depending on whether this unification can arrive at the technology for production, it will become either the starting point or the end-point of self-reliance in technology. In the Japanese experience, first design technology that was easily transferable was mastered, then conceptual design, leaving the design of details to foreign manufacturers. And while mastering the assembly of delivered machines, Japan began independent production of copied component parts.

The most difficult technology to change from user to manufacturer was the locomotive. In terms of the five stages, the gap between the fourth and the fifth stages was enormous. Japanese engineers dealt with the problem as follows.

In 1910, Japan ordered 12 locomotives from Germany, 26 from the United States, and 12 from Great Britain. The basic specifications to be met were: axis location 2C, weight 52 tons; axis weight 12.5 tons; diameter of driving wheel 1,600 millimetres; cylinder 470 x 610 millimetres; pressure 12.7 kg/sq m; whole conduction area 159.8 sq m; heating area 28.5 sq m; height of central line of the can 2,286 millimetres; valve equipment Fullshart type; superheater Schmidt type; maximum speed 96 km/h.

When the first of these new, powerful locomotives was delivered, Japan National Railway engineers and manufacturers carefully measured and tested each part and were able to produce an exact copy using the same materials.

Reflecting the technological situation of each country, the same locomotive could not be manufactured from the same specifications. For example, those made in Great Britain were not equipped with the superheater because the country lacked the technology for it. Perhaps it was to be able to compare the differing models that accounts for Japan's purchasing locomotives that were at variance from the original specifications. In any event, through the dismantling, reassembling, analysis, and testing of each component part, the Japanese engineers gained a detailed knowledge of the design and manufacture of locomotives from around the world. The different national models provided a good opportunity for Japanese manufacturers and repair houses to make a careful comparison.

Through this careful study and testing, Japanese engineers mastered technological know-how from design to production and were able to adjust designs according to the limitations of manufacturing. Thus, in the following year they were successful in achieving improvements to increase the boiler and water-tank capacities.

By the first quarter of the twentieth century, completely new types of locomotives for both passenger and freight trains were being manufactured, based on designs to fit the natural and technological conditions of Japan. These were the C51 locomotive for express passenger trains (1919) and the D50 for freight trains (1923), both enjoying a reputation for outstanding design.

In this way, the technological background was prepared for Japan's entrance into the golden age of railways in the 1920s. It had even become possible to attain speeds on narrow gauge tracks that traditionally could be obtained only on wide gauge.

Electric locomotives appeared on the scene around this time, and the Japan National Railway developed its own power sources and promoted electrification. Although various types of electric locomotives were imported from abroad, they brought with them many troubles. The operating conditions in Japan were quite different from those in the countries where they were manufactured, the level of technological stability was low, and there was a lack of interchangeability of component parts. A policy of domestic production was promoted as a result, and in 1928 Japan produced the EF52.

The first locomotive to be manufactured in Japan was built in 1893. Using a foreign design and foreign expertise, it took two years to build what can only be called a handcrafted locomotive, and without benefit of a pre-existent base of machine manufacture or heavy industry. This is a good example of how, with a model and many skilled workers, it is possible to produce a product.

Such technological independence was pre-modern independence, in essence, and had strictly a cumulative nature; it had neither national linkage nor international competitiveness. The product—the details and mechanisms of which differed one from the other—could be repaired only at the place of manufacture. Consequently, the products were not industrial products so much as examples of handicraft. In this sense, self-reliance in railway technology was not actually achieved until the years 1910–1920.

In any case, the example of the locomotive, and others as well, demonstrate that the ability to imitate is an important element of technology transfer and development. And yet, needless to say, in imitation, the level and quality of production will differ according to the varying technological conditions. The circuit from imitation to original production is likely to be repeated in many countries.

The Role of Foreign Engineers

As already stated, foreign engineers did not play positive roles in iron manufacturing. But the situation was different regarding railways. It was fortunate that the industry was blessed with such talented men as Edmond Morell (1841–1871), who came to Japan in 1870 as the chief engineer for the Keihin Railway Co., and William F. Potter, construction engineer. Morell opposed the importation of ties made of iron because those purchased earlier had proved unsuitable for the humid Japanese alluvial soil and because good-quality wood for ties was available in Japan.

Morell also proposed that the government establish an independent department to specialize in the administration of construction, including railways, and that the department should be in charge of cultivating technology experts. His recommendation led to the establishment of the Ministry of Works (1870, abolished in 1885) and, with the support of enlightened engineers, to the 1871 establishment of an engineering school, the Kogakuryo (this eventually became the Engineering College and later the Engineering Faculty of Tokyo University). Morell, who is buried in the cemetery for foreigners in Yokohama and hailed as "the benefactor of the Japanese railway," laid a solid foundation for this industry during his few years in Japan.

Morell and Potter were highly perceptive of the special characteristics of the Japanese economy and culture and thus knew how to evaluate and effectively apply traditional Japanese technology and skills. The presence of many excellent foreign engineers contributed to Japanese know-how in bridge construction and tunnel excavation and to a growing sense of confidence and expertise on the part of Japanese engineers. Thus, the successful transfer of a technology involves the cultivation of technological talent among the recipients of the technology.

The 30-kilometre connecting line between Shimbashi in Tokyo and Yokohama opened in 1872; the one-way journey required 50 minutes, and five round-trip shuttles were made each day. The advent of rail travel introduced a new equality into Japanese society: on the trains, regardless of social ranking, everyone rode together. The system of ranking by fares was established at a later stage, when the private rail comparies introduced three ranks of cars—white, blue, and red.

By 1982, the Japanese National Railway had become a gigantic modern enterprise with 21,400 kilometres of total operating distance; 6,944 million passengers annually; 100 million tons of freight; 5,290 stations; 3,757 locomotives; 18,235 trains; 4,683 diesel-powered railcars; 556 passenger trains; 84,923 freight cars; and 387,000 employees.

On the other hand, around this time the enterprise had been accumulating an average annual deficit of ¥1,461,900 million (US$6 billion at the rate of 240 to US$1), and its accumulated deficit up to 1982 amounted to ¥18 trillion (US$108 billion), and thus it was forced to pay an immense amount of interest. Nevertheless, the national railway was carrying 40 per cent of all passengers (9.7 per cent of all freight), and one can thus see that the role of the railways in Japan's modernization has not ended, although the industry is no longer the leading sector of technology.

10

Technology for the Textile Industry

The Textile Industry's Place in the History of Japanese Technology

The textile industry had a decisive influence on and was a leader in the economy and technology of modern Japan. Much has been written on this, and I have benefited greatly from the work of previous scholars. Of interest here, however, is the industry's relation to development and technology transfer.

Characteristics of Technology Transfer

In contrast to the shipbuilding, iron manfacturing, and railway industries, which developed in response to the "needs of government," the technology of the textile industry was developed to meet the "needs of the people." Specifically, the industry developed by supplying good-quality cotton cloth at low prices to people who had been "half naked," according to the record of a British traveller in Japan immediately after the opening of the country. This development was accompanied by a revolution in materials, wherein linen was replaced by cotton, a change that was reflected in the dyeing process as well. Besides dark blue and brown, woven fabrics of striped patterns using dyed yarns were produced, and these diversified and enriched the market for cloth.

Clothing was expensive and therefore used and mended carefully. Old bits were stitched together to form new pieces, and quilted waterproof winter coats were made by sewing with thick threads so many times that the original piece could no longer be discerned. Silk clothes were passed down for several generations and could be sold for a good price if the need arose. There was a big demand for used clothing; indeed, each entertainment district in today's Tokyo previously had a market for used clothing, and a used-clothing network covered the country. The markets and routes of those days have

130

become the distribution bases and routes for today's ready-made apparel (Nakagome 1982). In short, the technological changes that took place upstream were smoothly channelled downstream to create fertile new fields of trade expansion.

Regarding production, only spinning, which demanded a high input of labour and skill, was mechanized and modernized, as power spinning was 20 times more efficient than manual spinning. Weaving, however, was left to traditional technology (Ishii 1986). Modern technology was transferred only in the spinning division partly because Western weaving machines could produce only broadcloth, while the demand in Japan was for narrow cloth. Thus, modern technology did not eliminate all traditional technology; rather, both shared interdependent or mutually supplementary relations, and, for these reasons, the textile industry and its technology became stable and could develop into a national industry.

The links established between domestic and foreign technologies and between old and new technologies are the most important factors for the formation of a national technology network. In the textile industry, the links between the old and the new expanded and deepened the network of factory, urban-centred, mechanized labour on the one hand and domestic, dispersed, manual, traditionally oriented labour on the other. Thus, foreign industry was transformed into native industry.

As mentioned, power spinning was far more efficient than manual spinning, but the switch was not made without a careful consideration of what technology would best suit the existing needs and conditions. The mule spinning machine was the first to be imported, but it was replaced with the ring spinning machine in less than 10 years. The ring machine was imported not because it was the most advanced but because it required fewer skills and could be operated by low-waged women workers. The mule machine required skilled manual labour, traditionally that of only adult males; and it took a long time to form such skills. How significant the skill-saving was could be seen in the fact that productivity in Japan even with the highly efficient ring spinning machine was only one-fourth to one-eighth the productivity of a skilled British worker operating a mule machine. Although it is dangerous to make simple comparisons between workers' skills, we estimate that there were differences in output of 12 to 20 times. This great difference led to widespread adoption of the ring machine.

It must also be pointed out that there was a big demand for the thick, low-count yarns that were suitable for the ring machine. But the ring did not completely displace the mule (this is discussed in more detail later), as the mule machine was suitable for thin, high-count yarns used for high-grade goods. Nevertheless, the ring machine took the lead in the industry and the technology, but, to the extent that it did not represent a displacement of older technology, a connection between old and new technologies was established to create an extensive market and, thus, a new national industry.

The important aspect of this development was that, rather than an unconditional or blind transfer of the most advanced technology, the selection of a technology was made based on rational judgement in terms of skill-formation costs and marketability.

The Role of Government

The Spinning Industry

The development of the spinning industry owed much to the efforts of private individuals, men like Shibusawa Eiichi (1840–1931), a characteristic peculiar to the spinning industry. Furthermore, although mining technology, for example, was also developed by private companies, the mines formed isolated "enclaves of compound technology," while the spinning industry was urban, with large-scale plants clustered around the Kansai area, especially in Osaka. The silk-reeling part of the textile industry operated on a smaller scale and was scattered over a broad area.

This difference between spinning and silk reeling was determined by the supply of raw materials. Silk reeling existed where the raw materials were produced because of the particular time and seasonal constraints associated with sericulture. The cotton-spinning industry, on the other hand, used imported materials, and so the factories were near ports and the centres of consumption. In both industries, the labour force was made up mainly of women. The two industries also differed in terms of basic operation; that is, the silk-reeling industry did not operate from winter to early spring, and workers were contracted by the year. In the cotton-spinning industry, a 24-hour, 2-shift (daytime and night-time), all-year system was in force, and the contracts were signed usually for three-year periods.

. Cotton was grown nearly throughout Japan, and, as a commodity and for domestic consumption, it was the most important crop after food crops. However, because of the diverse cropping conditions of each region, cotton quality was not uniform, nor was cotton suitable as a raw material for machine spinning, a main reason why imported cotton was used. As a result, Japanese farmers lost an important income source, which, in turn, contributed in its way to the conditions whereby the spinning industry employed farmers' daughers for work in the mills. In the southern part of Osaka, which was a major cotton-growing area, farmers attempted to cope with this difficult situation by means of producing shell buttons, hairbrushes, and later, eyeglasses and miniature lamps for Christmas, all of which were mainly for export.[41]

Despite such differences between the spinning and silk-reeling industries, initial technology transfer was made by the government in both cases. The government-operated model spinning and silk-reeling plants were, however, short-lived. Both were poorly managed, and a fiscal crisis of the new government was an impetus for their sale to the private sector.

Where the government succeeded was in the transfer and diffusion of technology in silk reeling. Although Japan was an underdeveloped country in silk-reeling technology, it accomplished the targeted goal of earning foreign currencies from the export of silk.

In the spinning industry, it was successful in substituting its own products for imports (which had accounted for 30 per cent of Japan's total imports) and later started to produce goods for export.

Nevertheless, this partial success was overshadowed by the government's failure in terms of business management. For example, in the spinning industry, the government set up 10 small plants (each with 2,000 spindles, that is, a set of 4 machines, each with 500 spindles) in various places in Japan. The 10 plants were part of an anticipated 255 plants that, if established throughout Japan, would make possible the complete replacement of imports with domestic products. These first 10 plants did not succeed, however, and among the causes for the failure, the following could be considered as the main contributing factors:

1. 2,000 spindles were too few to cover costs, but the government could not afford to make additional investment to attain economy of scale.
2. Because the power source was the water-wheel, spinning had to compete with rice cultivation for water, and farmers had a vested priority. During the winter, free from this competition, the spinning industry was nevertheless plagued by a shortage of water, which prevented it from operating continuously throughout the year. As an alternative, some plants adopted wood-burning steam-engines, but wood was not always available. In the end, the power problem (and the technology to use power) went unsolved.
3. Because no one was familiar with the machinery, the machines were not installed properly, and they often vibrated (according to a report from the Mie spinning plant). Among other problems, the vibrating led to defective goods, and, as engineers were in short supply, machine repair and maintenance were neglected, which resulted in extremely poor production. The idea arose to employ foreign engineers, but none of the plants could afford to.

Referring to a specific case, at the Mie spining plant, the fly frames and the spinning machines were manufactured by different makers, and the speed of work on the two was not balanced. This fundamental lack of planning coordination resembles the current situation in many developing countries.

The total number of plants established by the government reached only 17, and when, after about 10 years, the government withdrew its support, only 3 managed to survive. These three evolved into the two large textile concerns of Toyobo and Kurashikibo. By working to overcome the technological weaknesses described above, they achieved a leading rank among rising new industries. These plants were also successful in management reform, which was directly connected with the resolution of the technological problems.

Despite the failure of the model plants, in 1881 a plant (the Osakabo plant) was established independently of the government plan by private financing after a carefully conducted feasibility survey of the location. The plant started with 15,000 spindles, using 150 hp steam-power. A youth named

Yamabe Kentaro, studying in England, was sent to a plant in Lancashire for training, while the men who would later become the leading engineers under him were sent to government plants. The machines were installed under the guidance and control of a man named Needle, a British engineer, and Saito Kozo, of the Osaka Mint Bureau. These two had learned much from a thorough study of the errors made at the government-operated plants.

One might ask at this point why 2,000 spindles were decided upon as the standard number at the government-operated plants. Previous studies do not provide a convincing answer. Presumably, it was impossible for the engineers involved in the project to conceive of a 10,000-spindle factory. They had not seen plant operations in any Western industrial country and thus probably lacked sufficient knowledge of the latest methods of engineering.

In short, the role of government in the spinning industry was not positive, and even if we admit its role, it should be regarded as limited.

The Case of Silk Reeling

As with the railways, a foreigner living in Japan (a Dutchman, or, according to some, a Frenchman) filed an application to build a silk-reeling plant. As a result, the government hastily decided to establish such a plant itself. Fortunately, it was able to hire Paul Brunner (1840–1908), who, over a long period, had earned respect as the "great master." Most of the country's technological success in silk reeling should be attributed to him.

For the Meiji government, there was only raw silk, besides its gold, silver, copper, and coal exports, that Japan could export to pay for its huge amount of imports. However, Japanese raw silk was produced with an emphasis on its white colour, lustre, and softness; its denier was inconsistent and so its commercial value in Europe was low, where it could be used only for woof in weaving. Because it was necessary to adopt Western reeling technology to produce hard twisted and uniform thin thread, the government imported plant facilities from France; French engineers and skilled workers were also hired to provide guidance. This was in 1872, and the location was a small castle town, Tomioka, in Gumma Prefecture.

In addition to French technology, Italian technology was also imported. But it took 20 years before these technologies, which had been introduced almost at the same time, could be cross-bred to form a system of technology most suitable for Japanese conditions. Once formed, Japanese reeling technology became stabilized and its market moved from Europe to the United States, where it could compete with low-priced Chinese and high-quality Italian silks. Its international position was thus secured.

Besides mastering the Western technology of reeling, the Japanese provided technological innovation by improving the unevenness of the thread through standardization of the cocoons and by achieving a thinner thread through pre-treatment. Furthermore, a series of improvements was made to the cultivation methods of mulberry and the raising of silkworms. In the

1930s, the technology was at its peak; then came World War II and the immediate onslaught of stagnation in the industry.

One significant aspect of the government operation at Tomioka was that it employed women from all over Japan and trained them—over a period of two to three years—in the entire production process. As a result of their extensive training, when they returned home, they became teachers and directly contributed to the diffusion and development of new technology throughout Japan.

Especially notable were the intelligence and great skilfulness of these workers. This is evident from the *Tomioka Diary*, written by one of the women workers at the plant, Wada Ei (1857–1929), in the days, it might be noted, before compulsory education. The diary indicates that the workers were highly knowledgeable, which, added to their skills in traditional technology, enabled them to master and begin to pass on the new technology in a matter of just a few years' time. Not insignificantly, many of the women at the Tomioka plant, it might be mentioned, came from former samurai families.

Second, as Wada wrote, in establishing the new technology in other areas, such drastic capital-saving measures as the substitution of wood for the copper, iron, and brass used at the Tomioka plant, wire for glass, and an earth floor for the brick were adopted. Investment in the Tomioka plant amounted to ¥200,000, equivalent to ¥2 billion today; a plant established at Matsushiro started with only ¥300.

The boiler used at Matsushiro had been designed by a sailor and was built by a copper-pot artisan; the pipes were made by a gunsmith, and the wooden gears by a maker of spear shafts. It was a full mobilization of traditional technology.

After much effort, machines were built and made operable, a stunning achievement. Despite that the equipment was a crude imitation and had poor output, the mere ability to manufacture this equipment revealed a formidable engineering ability among the workers. This means that the technological gap between Japan and the advanced countries at the time (disregarding the principles of engineering) was not desperately wide. And yet, to many of those who thronged to the Matsushiro plant to learn how to set up similar plants in their towns and villages, even the relatively crude hand-crafted machines at Matsushiro appeared needlessly luxurious.

The example of Matsushiro makes evident that management in the silk-reeling industry followed rationalistic management principles, with a keen sense of evaluation of technology. The managers were practical leaders, who worked on the shop-floor designing and manufacturing machines of their own and enlarging the scale of operations by repeatedly investing their profits. This was in sharp contrast to the cotton-spinning industry, in which many of the managers were politically connected businessmen.

On the other hand, management in the silk-reeling plants did not hesitate to coerce the young female workers in to accepting their ethic of self-sacrificing diligence. In the plants, a wage system was worked out according to rank, in which the lump-sum amount of wages was fixed and workers had to compete to take a greater piece of the pie. Based on their abilities to conserve raw materials and to reel a better-quality yarn in greater quantities, the workers were ranked first or second class, and wages were determined by these rankings (Nakamura 1952).

Regarding the government's role in the technology for the silk-reeling industry, it was important, but only for a short period at the very beginning, the initiative for development being held by private business groups. Legislative measures against the harsh working conditions and hygienic problems always lagged, and in the administration of the law, a higher priority was put on protecting private property than on protecting basic human rights. In labour disputes, the government authorities continually opposed labour.

How Japan Was Able to Catch up with India—A Subject for Dialogue

For the Japanese cotton industry, India was a stiff competitor and a technologically more advanced country. Here the issue concerns not only the formation of international competition around a technology but also its influence on decisions regarding the priorities in development policies. Some advocates give first priority to light industry, especially the textile industry. Our position is that no industrial policy is equally applicable to every country, and so the discussion here is meant only to further our dialogue without imposing hard and fast conclusions.

Management Scale and Technology

A simple comparison of Japan and India reveals there was no restriction on expansion and development in the Japanese cotton industry, in contrast with the situation in Great Britain and India (Yonekawa 1981). There were only economic limitations. Thus, for example, in the Japanese cotton industry, there was a 24-hour, 2-shift (day and night shifts) operation system in effect, and the high profits earned from this system were reinvested. Before 1911, no legislation existed to protect the workers.

This, then, enabled Japanese enterprises, the initial scale of which was much smaller than that of the spinning shops in Bombay, to catch up rapidly.[42]

According to Kato (1979), a shift from the mule to the ring spinning machine was carried out simultaneously in Japan in all large factories, contrary to the situations in England and India, where both types were in use. In India, there was the big factory at Tata that converted to the ring at an early stage, but in India as a whole, the mule spinning machine was in the majority and the ring machine was auxiliary.

The question regarding types of spinning machine relates to the problem of management. Tata was exceptional because, from the beginning, it was opposed to the proxy management system (foreign-owned, locally managed enterprise system); whether we may point to that as the reason for Tata's early conversion to the ring spinning machine is uncertain, but it does appear that India's proxy system by nature was passive toward technological change. Indeed, the switch to the ring spinning machine in India was slow, and, in comparison to owner-managed operations, the proxy system was not sensitive to technological change.

On the other hand, when Japan was making frantic late-comer efforts to master the new technology to replace imports with domestic products, because (1) the ring spinning machine required fewer skills to operate and (2) domestic demand was mainly for low-count yarns, the ring spinning machine was most suitable: it was not only the most advanced, it was also the best suited to the market and working conditions in Japan at the time.

When the large-scale factories, specialists in spinning low-count yarns, started also to produce woven fabrics—at the time of their entry into the Korean and Chinese markets—a development in the technology of cotton mixing was introduced; this change was accomplished through minor operational improvements in the functioning of the ring machine for spinning middle- and high-count threads and for weaving. The technology of mixing had thus developed into a Japanese specialty.

Trading Companies and the Technology of Cotton Mixing

While the technology for cotton mixing was also being applied in other countries (especially India), Japan, which could not produce high-grade raw materials, had originally developed this technology and established it as a technology for making use of inexpensive cottons. Table 3 illustrates the process of spinning and weaving high-grade fabrics by changing the ratio of the cotton mixture, depending on the characteristics of the ring spinning machine.

From the table one can infer why it was that in a few years' time the cotton-spinning industry stopped using Japanese cotton except in producing goods that made the best use of the particular nature of Japanese cotton for dyeing.

The development of cotton mixing had another aspect, the links with other industries.

Unlike the cotton spinners in Lancashire, who had no need for business organizations or related activities for purchasing raw cotton and who could concentrate all their energies simply on spinning, in Japan—and this was so up until the Second World War—there was a close relationship between the trading companies and the cotton industry; this relationship functioned in such areas as the import of raw cotton, credits, and in sales of manufactured cotton yarn and woven fabrics (Kato 1979).

A relationship involving specially contracted spinning between trading firms and manufacturers developed. In this arrangement, the trading firm supplied a certain volume of raw cotton that it secured through direct pur-

Table 3. Fabric grade and the ratio of cotton mixture

Counts	Japanese middle grade	Cotton of other countries
−10	Almost all	Almost none
10–13	30% to 70%	30% to 50% (Indian and Indonesian cottons)
14–18	30% to 50%	50% to 70% (same as above)
20–24	10% to 20%	80% to 90% (Indian high grade, Indonesian and American middle grade)
28–		100% (higher than American middle grade)
Over 50		100% (American high grade)
Over 60		100% (Egyptian low grade)
Over 80		100% (Egyptian high grade)

Source: From Sambe Sei'ichiro, ed., *Meiji shoki ni okeru wagakuni menka seisan no chōraku* (Decline of cotton production in Japan at the beginning of Meiji) (Tokyo, 1947).

chase at the place of production, and sold the manufactured goods. This sort of relationship between large trading firms and large spinning factories typified the industry, and here we see the prototype of the function of a Japanese trading firm. There were also secondary and tertiary linkages, typified, for example, by ships engaged in the transportation of raw cotton from all over the world.

Subdivision and Rotation in the Production Process

Regarding the production process, many specialists stress the 24-hour operation of the cotton spinning mills as important, but I contend that skill formation through the rational subdivision of production work and personnel rotation was just as important.

Although the massive employment of unskilled workers and long hours defined factory labour at the early stage of industrial development, the tasks were simple, and versatile, skilled workers did not come about under these circumstances alone. The technology of cotton mixing entailed operating machines that required skills, and, with woven fabrics, a higher quality required a higher degree of skills.

Consequently, management inevitably adopted measures for the development and formation of skills. One such measure was the rational breakdown of the major production processes into several subdivisions, which then made it easier to quickly master the skills at each subdivision. Workers moved from one subdivision to another, which enabled them to master the technological knowledge and skills needed for each major production process. This was the Japanese method of skill formation.

Hosoi Wakizo (?–1925), who left a detailed record of the life of textile workers entitled *Joko Aishi* (The sad story of women workers, 1925), and who worked the reel in mule spinning, wrote that around the time when the Japanese textile industry established its international position, there was a dual structure in wages and technology corresponding to plant size and operational scale. Moreover, there were important differences in the type and scale of operation between exclusively spinning plants and spinning and weaving plants. Hosoi pointed out that workers in large and small plants had lost the interchangeability economic theory requires as a premise. Hosoi himself was a single-skilled worker in a large-scale plant. He described the production process as in the chart on pp. 140–141.

This description pertains only to the spinning division as Hosoi saw it. Actually, there was a much greater division of work. If the division in weaving (8 major processes, 18 subprocesses, and 18 different types of jobs), their auxiliary processes, and the chief supporting divisions (motors, repair, construction, maintenance) were added to those above, the entire plant consisted of 50 to 70 types of jobs (or subprocesses).

Part of the chart came from a report entitled *Nihon no kasō shakai* (Lower strata of Japanese society), by Yokoyama Gennosuke, published more than a quarter-century before Hosoi's report; the major divisions of Yokoyama's chart are marked in the above chart with asterisks to clarify the difference between the two. Yokoyama did not touch on the auxiliary divisions (or the weaving division). Although Yokoyama and Hosoi were describing two different plants, a comparison of the two charts indicates that the technological organization in large plants had been standardized by the end of the nineteenth century. Presuming that what Yokoyama referred to as the "ring" indeed was the ring spinning machine, and considering that the difference of about 30 years between Yokoyama's book and Hosoi's report was the difference between the initial stage of spinning technology and the stage of its maturity, it may be supposed that, while the ring machine itself had undergone some improvement, there were no revolutionary changes in the technology.

Toyoda Sakichi's automatic power loom was invented in 1925, the year in which Hosoi completed his report. But Toyoda's loom did not immediately become popular: In the smaller factories, there was a lack of capital for introducing the new machines, while the larger plants felt, as yet, no need for them. A sophisticated division of labour in the production process and the use of specialized (i.e., single-skill) workers had reached its development in the large factories. At small plants, according to Hosoi's technological classification, one worker was still in charge of the whole range of tasks in one division, which presumably corresponds to our implied "major process." The system of division of labour at small plants was a vestige of the mule-spinning stage.

The Osaka Spinning Co. introduced the ring spinning machine in 1886 when its second plant started operation. Major plants followed Osaka's lead, but, as Hosoi confirms, these major plants did not abandon the mule spinning machines, whose operation required a long time for workers to master. The

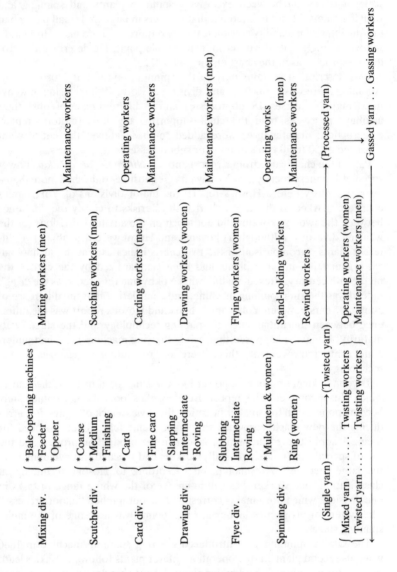

Reeling place   Reeling workers (women)   Rewinding worker

Rewind   { Rewinding

Bundling   Reel-tightening
workers (women)

*Packing   Stretching workers

The process up to the above was the actual production process. There were also the following auxiliary divisions:

Roller shop   { Roller repair workers (men)
Roller polishing workers (women)
Belt workers

Experiment div.   { Grain testers (men)
Assistants (women)

Cotton-selecting div.   Cotton-selecting workers (women)

Lubrication (men)

Bundling (men)

Bank making (women)

Transportation (men)

mule was preserved probably because of its special mechanical characteristics relating to the production of high-count yarns and fabrics made from long-fibre cottons.

At this stage, the competitor was no longer India, which manufactured low- and medium-count yarns, but Great Britain. There, a high degree of horizontal division of labour had been developed among the three divisions of spinning, dyeing, and weaving, and each plant had become smaller, and few plants were engaged in both spinning and weaving. Japan was characterized by a vertical integration on the basis of large-scale operation.

We are not in a position to judge whether a horizontal division of labour is more effective or whether vertical integration is more rational in the process from technological dependence to technological self-reliance (especially technology development).

However, we can offer an interesting example of the value of the latter approach: the success of the Japanese watch industry in moving from quartz to electronically based watches was due to maintaining a continuous production system in the plants. Considering that the principle of the quartz watch had been discovered more than 50 years ago, the opportunity to apply it was equally available to all watch manufacturers. However, in Switzerland, for instance, where the famous watchmakers have specialized in design and assembly, the mechanical watch industry has been based on a system made up of a wide range of independent, highly skilled manufacturers of component parts. As a result, the Swiss watch industry, in consequence but not in principle the most advanced in the world, was not able on its own to lead the innovation in this sector of technology.

In textile technology, as Hosoi has recorded, the textile workers undertook "50 to 70 different types of jobs," and actually even more divisions of labour. Yet it was a general practice at large plants to group the workers into "operating workers" and "maintenance workers," although it was diffcult to distinguish clearly between the two. Eventually, a precise division of labour came into force, but it was not fixed by job functions. This is noteworthy when compared with India at the end of the nineteenth century.

Technological Change and Female Labour

The conversion from the mule to the ring machine changed the labour structure from one dominated by male workers to one with female workers in the majority. The ring spinning machine required no physical strength, and wages for women were lower.

According to Yokoyama, female workers were now performing the tasks of drawing, slubbing, spinning, and reeling. The spinning division needed the greatest number of workers, and the division where the work was hardest was slubbing. The spinning industry developed subsequently depending on women workers. This industry had the potential to create a huge employment opportunity on the one hand, but, on the other, it depended on the possibility of employing massive numbers of low-waged workers. To this end, it often located in urban areas in places near where the poor and the indigent lived. The low price of land was perhaps a factor, but the possibility of recruiting a cheap labour force was undeniably of primary concern.

Because of the low wages, often not only male workers but also their wives and young children were employed together; the men working alone was not enough to provide families the necessary support. This employment of all family members is referred to as "complete employment," in ironic comparison with the term "full employment." (These terms are said to have been coined by Professor Tohata Seiichi.)

The gruelling work and long hours at the factories caused high turnover, and recruiting additional or replacement labour from the area where a factory had been established became almost impossible once several years had passed since the establishment of a factory. Consequently, when recruitment of labour had approached its limit, companies would customarily dispatch agents in charge of recruitment to remote areas where people were likely to be unfamiliar with the reality of the shop-floor. Using three-year labour contracts and low advance money, the companies got a firm grip on their workers, most of whom, as mentioned, were women. More than 80 per cent of the labour force in all factories was recruited in this way. Urban applicants accounted for the rest. Women comprised more than 70 per cent of the total work-force.

Such was the employment structure in the textile industry. Since more than 30 per cent of women workers retired every year at the expiration of their contracts, it was necessary to replace them annually.[43] The expense of recruitment amounted to several months' wages, which the newly recruited women workers had to bear, and it was thus the same in effect as a reduction in wages. In the absence of recruitment expenses, the wage would be high by the amount thus saved; but, even this higher wage was insufficient to secure applicants in the number required.

An overwhelming majority of recruited women workers were the daughters of families engaged in agriculture or fishery who were forced to seek employment to help support their families. As a result of the terms of their contracts and their meagre advances, they were forced to live in dormitories, where they were under strict control; to prevent them from running away or being hired by another company, they could almost never leave the dormitory. Every two workers alternated shifts, one day and the other night, using the same bed continuously; this gives an idea of how harsh the conditions were.

Such conditions did not exist at all of the mills; conditions were less harsh at the Tomioka dormitory, for example; nevertheless, Tomioka had many sick women, and the Tomioka mill was the first factory in Japan to have a hospital attached to it. Women in silk-reeling work could, in the early years, recover at home during the winter season (later the situation deteriorated to the poorer level characteristic of cotton spinning).[44] Despite the development of the spinning industry and technological change, its female labour force did not benefit from improved conditions.

Hosoi claims that what tormented them most were the long, hard hours and homesickness. Their hard work at home in the fields did not prepare them for the harsh conditions of the mills, for the severe factory regime they encountered there. Therefore, despite the attempts by the companies to keep them on, many women workers returned home, often with their health irrevocably ruined.

As a result of the bad working conditions, long hours, and poor nutrition, beriberi and tuberculosis were not uncommon. Many of the women suffered from impaired hearing caused by factory noise. Their tendency to speak loudly as a result frequently brought on ridicule. Eventually a cure was found for beriberi; and, once its cause was discovered, the factories began giving the women workers a more nutritious—albeit less easily digestible—type of rice. And because they could not rest after meals, many women developed stomach trouble.

Through infected workers who returned home, tuberculosis spread so widely that it was called the national disease. According to Hosoi, the birth rate among married women workers was generally low, and the incidence of miscarriage and premature births among them was more than twice the national average; the number of births of handicapped children was also high. There was no protection for pregnant women, and the sole condition for employing juveniles was a period of compulsory education. Each factory overcame this by establishing a school attached to the factory. However, this educational component only added to the already high degree of overwork.

The Starting Point of Development

With the cotton yarn and silk thread spun by women workers at the sacrifice of their health, Japan bought warships, purchased machines, increased exports, and substituted imports with domestic goods. It achieved technological independence and survived as a sovereign nation. And for the sake of securing a "greater co-prosperity sphere," Japan ravaged Asia and damaged itself seriously. The price paid by the workers was too high.

Japanese self-reliance in technology and development after World War II was initiated with the recovery of pre-war technological development. And now, the high price that had once been paid by women workers was, after the war, paid by the Japanese people as a whole through environmental pollution.

The sad pre-war history of women workers and the post-war environmental pollution occurred because the élites of politics, administration, economy, and technology failed to concern themselves with the consequences of their arrogantly enforced "development," which cared little for the rights of ordinary people; the sole aim was urgent national development. As far as it pertains to human rights, the relation between the urgency and supremacy of development and also between development and democracy may need to be made open to criticism by foreign citizens, even though a country's development is a matter of its sovereignty. Otherwise, the international basis of our "methodological dialogue" will have no meaning.

Before closing this section, a brief mention of the contemporary Japanese textile industry is in order. The Japanese textile industry today no longer depends on natural fibres. Although natural fibres are useful and sometimes indispensable, they have lost the importance they had before the war. Hand-woven linen, cotton, and silk cloth made of only natural fibres have become extremely high-calibre craft work.

Most of the leading textile companies in Japan today have been engaged in the research and development of chemical fibres since World War II, when Japan could not import raw materials, and have thus diversified into the chemical industry, as well as into other sectors of new technology. Japan's technological stock has prepared it to enter the most advanced fields of technology; indeed, although most Japanese are not aware of this, vinylon (polyvinyl alcohol fibre) was invented in Japan. Furthermore, the trading companies that specialized in raw cotton and cotton products have become general trading companies, serving technology transfer as a part of their activities.

11

The Transformation of Small-scale Industry into Modern Indigenous Industry

Small-scale Industries in Development

A number of large-scale factories (with more than 1,000 employees) in Japan have shut down since the second oil crisis, and this can be attributed to technological innovation. The need to save energy and resources has had an effect in the areas of labour, materials usage, and energy consumption. A fierce competition to continually come up with a new product has emerged and has led to notably shortened life spans of goods. These developments have resulted in a decrease in the number of large-scale factories. High consumption levels have transformed the mass-production system into a system producing high-quality goods in small quantities to meet market needs and to diversify risks. Under these circumstances, the traditional Japanese employment pattern has been eroded, some of the effects of which have been mentioned in the preceding section on female labour.

The size and importance of the role of medium- and small-scale industry in the whole of the Japanese manufacturing industry is not widely known in the third world. Neither is it known that there is a structure linking these industries with the more internationally famous Japanese enterprises in business and technology.

The definition of medium- and small-scale industry has differed according to the period, varying in maximum complement from 10 to 20 to 100 employees. Today, government classification designates enterprises with less than 300 employees and capital of less than ¥100 million as medium- to small-scale. [45]

According to statistics, factories with fewer than 20 employees account for 87.3 per cent of the total number in Japan, employ 20.1 per cent of all workers, and contribute 12.6 per cent of the total national output.

Factories with more than 500 employees, on the other hand, comprise only 0.3 per cent (1,807 total) of all factories in Japan; they employ 20.5 per cent

146

of the nation's workers (2,246,000) and account for 38.3 per cent of total output. While in Japan factories with fewer than 100 workers make up 98.0 per cent of the total and employ 58.0 per cent of all workers, in the United States, the respective figures are 87.7 per cent and 25.4 per cent, and in West Germany the corresponding proportions are 72.6 per cent and 18.7 per cent. The percentages for factories in Japan employing more than 1,000 workers are 0.1 per cent and 13.4 per cent, in the United States 0.6 per cent and 27.5 per cent, and in West Germany 2.2 per cent and 38.0 per cent.[46]

Aside from the statistical significance of these comparisons, it is clear that even in highly industrialized countries, medium- and small-scale factories have a role, and that, depending on the type of technology and industry, an enlargement of scale may be unwise or impossible.

Japanese medium- and small-scale enterprises were forced to renew their equipment in search of high efficiency as they faced a serious shortage of labour during and after the rapid economic growth of the 1960s. The two oil crises forced them to confront increased costs in both labour and materials. The changes and intensification in competition forced them to renovate their operations. Some of Japan's famous enterprises that maintained a small scale as an ideal size for the development of new products also underwent this process of adjustment.

From around 1975, the upgrading of facilities by small and medium enterprises brought about a new phase. The attainment of a high technological level has given the exports of these enterprises a competitiveness in international markets. The use of ICs in the production process has minimized differences in manufacturing capability and in the quality of products among manufacturers, so that the original equipment manufacturer (OEM) system has spread rapidly to enterprises of all sizes, small, medium, and large. Whether this represents a new stage of internal structure in the national network of technology is uncertain, but we may say it is a new phase, inasmuch as in the manufacturing industry, there have always been two opposing types, one seeking stability, the other continuous growth.[47]

The need in Japan for small and medium enterprises and their significance in society will not likely change. A good example of the trend is the fact that factories with fewer than 300 workers account for 99.5 per cent of the factories in Tokyo and employ 74 per cent of all factory workers there. Also of note is that small, medium, and large factories are located strategically, in accordance with the vital technological and business relationships they share.

In terms of development, what this process represents is the dissemination and development of modern urban industrial technology. In effect, the process is one in which those who have mastered the technology of a production process (or kind of job) at a specific level have separated it from the mainstream and become independent entrepreneurs (i.e. from process subdivision to process separation).

Providing an entrepreneur has a clientele, it is his technological ability that assures his independence. However, if the separation is made merely in the form of a change in the place of production as simply an extension of the subdivision of the production process, the new establishment represents in fact an affiliate of the parent company, much like a subcontractor within the plant. Furthermore, in some cases, depending on the type of industry and general business conditions, it will become necessary to master the technology of the entire production process to make the separation. In establishing independence, technology is transferred from the head shop, much as skills and knowledge are handed down from a master craftsman to his apprentice or, in the Japanese custom of *norenwake* (giving the name of one's shop to a former employee), one merchant helps another set up a business. This is easier to do with technology that needs little start-up investment (most such technology usually requires higher skills).

If the amount of initial investment is large, it becomes necessary to depend on borrowed capital, especially commercial financing. When this happens and materials and machines are leased, customarily the business starts as a processor and operates under a processing-fee system. As long as production is divided into separate small production processes, the processing fees remain low. Under these circumstances, the differences of skills, that is, the technologies of small independent enterprises, determine the differences in efficiency of production and of the use of raw materials. Many owners of small- and medium-scale enterprises are self-made men who accumulated technology and forged ahead on the road to self-reliance.

In addition to the classification of industry in terms of scale, it can be classified according to modern vs. traditional. Applying this classification, what one discovers is that most small-scale enterprises are in traditional industries and engaged in the production of consumer goods and services. While factory production uses modern technology, native industry depends on traditional technology, machines, and tools. In terms of scale, the range is from several workers to several hundred, and yet, according to one study, even in the 1930s, traditional industry output occupied a quarter of the gross industrial production.

According to statistics since the middle of the past century, 80 per cent of gross national expenditure has been for personal consumption and most of it for the consumption of traditional goods (foods, clothing, textiles, china ware, and other general merchandise). The position of small-scale industry in the national economy has been highest after agriculture.

As stated in regard to textile technology, yarn was manufactured at modern factories, while fabrics were woven in the traditional manner and places of production and sold through the historical wholesale system. Thus, the two were not in an exclusive relationship, but in a mutually supplementary, interdependent relationship, which aided the development of both. After World War II, the modernization of traditional technology changed this situation, and the scale of enterprise began to reflect the specialization in technology, though not without exception.

What is important is the formation of an interlinkage between traditional and new technologies by which traditional technology is finally modernized. It is the transition from a stage in which technology determines management to one in which management decides the orientation and level of the technology.

For this reason, the process of technological improvement is characterized by integration of management ability and the potential of the technology. The smaller the enterprise, the more it depends on management's technological ability.

It is noteworthy that, as early as 1900, before Japanese technology became self-reliant, the products of small industries made up a high percentage of Japan's exports. Raw silk accounted for 22.3 per cent, woven silk 9.3 per cent, green tea 4.0 per cent, matches 2.9 per cent, and silk handkerchiefs 2.2 per cent; thus, manufacturers using traditional technology accounted for more than 40 per cent of total exports. The last stage in match production (i.e. packing) depended on people working at home and was so labour-intensive that even young children were used among the urban poor, especially in the large cities, notably Tokyo and Osaka.

The Village Button Industry

Taking the case of the button industry might surprise readers, but it is worthy of consideration because it represents rather an unusual case but also one that illustrates very well the diffusion of small-scale technology. The discovery in developing countries of similarly specialized technological areas could have important consequences for economic and technological development.

Buttons began to be imported around the time Western-style uniforms gained currency with the military, railways, the police departments, and similar groups. Besides buttons made of metal and bone, those made from shells, which were used for underwear, were also imported, although Japan had abundant raw materials of good quality.

Button manufacturing on a modest scale started in Japan from around 1878. They were expensive but of high quality, made by metal workers (such as goldsmiths and silversmiths) using files, whetstones, and punches. The market was extremely small, however, because of the continued dominance of traditional Japanese wear, and, in order to establish a stable industry, button makers were compelled to turn to exporting their product.

Having noticed the existence of rich raw materials in Japan, a German named Vinkerel opened a factory, complete with an array of machines for making buttons, at the concession in Kobe in 1890. He was supplying the Japanese market with "German-made buttons" manufactured in Kobe. The bleaching process used in button manufacturing had been kept confidential by the German engineers, but a Japanese processor's solution to the problem of bleaching gave Japanese buttons, which had been treated as semi-finished goods, an advantage, thus forcing the German factory out of business.

The answer to the question of how this was possible is to be found in the thorough division of the production process. Production at the factory was broken down into more than two dozen microprocesses, each of which became a separate job performed by a worker "manufacturing" at home. Moreover, as far as possible, no machines or equipment were used except those traditional tools and methods that demanded little in terms of skill. The next step was to reduce the processing costs to an extremely low level. This was identical to the business control exercised by the merchant over his scattered manufacturers. The button manufacturers unified and managed their individual microprocessors in the same way.

The simplification of work and the low processing costs did not lead to the independence of the microprocesses, but rather promoted side jobs at home. What originally had been a modern urban industry was transformed into an industry that depended on the labour of lower-class urban citizens working at home. It then penetrated into suburban agricultural areas in search of cheaper labour; the target was enlarged from the urban informal job class to the rural informal job class. Although, in order to master the whole technology for a basic production process (or several major processes), as opposed to a single, small process, it was necessary that the worker become an apprentice of a "manufacturer," those who mastered the technology presented little threat of breaking away and becoming independent manufacturers, as their products were component parts rather than finished commodities.

The enlarged production of shell buttons brought profit to the merchant manufacturers. The shortage of raw materials caused by greater production and the conversion to and dependence on imported materials changed this situation, however. The sharp fluctuations in the price of raw materials brought on speculation and hampered distribution. When to this was added an increase in demand resulting from an economic boom, wholesalers and manufacturers were no longer able to undertake strict inspection of goods, and, as a result, the mass production of inferior-quality goods started. Holding down processing fees to too low a level can lead to this sort of situation.

In general, since the Meiji Restoration had done away with certain business restrictions, the problem of the mass production of inferior-quality goods was seen in almost all the traditional industries and technologies. The situation was the same for new technologies that had been transformed into traditional-type technologies. When the change of raw materials occurred, that is, the addition and development of new technologies, the old structure of the business world had to be reformed. To protect the common interests in each sector of business, the master-apprentice system of control had to be transformed—democratized—into a system of control by an association (see Takeuchi 1979; Takechi 1979).

The areas where rural industrialization developed were those in which commercialization of agricultural production was advanced. For example, the cases taken up in this study were in the southern part of Osaka and Nara prefectures, where cotton-growing and food-oil production had been active. The development of a modern cotton-spinning industry brought about the substitution of locally produced cotton with cotton imported from India, and the development of a modern food-oil industry, centred in urban areas, ruined the traditional oil-making industry in this area (Sasama 1981).

Kagawa Prefecture was another area where the button industry developed as a cottage industry after the traditional salt making (by the salt-field method) and sericulture lost their viability. The transition from the traditional salt-making and sericulture industries to modern button manufacturing was possible because of the long experience with producing for a broad market (Yasuoka 1981).

The successful and lucrative export of Christmas lights by farmers in Kagawa Prefecture during the chaotic period immediately after World War II is another example of the sort of adaptability that made possible the successful transition to new industries as the old ones lost their viability.

Similarly interesting accounts exist concerning other sectors, such as the clock industry, which developed in parallel with sericulture and which contributed to establishing the habit of punctuality among farmers, an important element in the foundation of modernization. Also important were the early glass industry (Kikuura 1979), the eyeglass industry (Ueda 1979), and the development of bicycle industry technology (Takeuchi 1979; Ueda 1980).

Regrettably, and surprisingly, the third world participants in our discussions showed no interest in the fact that an industry like the bicycle industry, though there are fewer component parts than in the watch industry, for example, could be used as a technological indicator if all parts are manufactured domestically, as was true in Japan. The complete domestic production of bicycles, when combined with engine technology, constituted the precondition for motor-cycle technology. The bicycle was useful as an index of technological convergence in Japan, and, as such, could be considered the first gateway for a newly industrializing country. High-quality ball-bearing manufacturing, in particular, is a good index.

Before moving onto the next section, it will perhaps be useful to elaborate briefly on this last point. Ball-bearing production is directly connected with the production of specialty steel, the material from which ball-bearings are made. A tremendous amount of technology must be accumulated before it is possible to manufacture bearings of a specific size with the same consistent quality. And the smaller the size, the higher the technology must be. Usually, it is not necessary to aim at such high-level production right off, and it is advantageous to depend on imports for precision-made, highly machined parts.

After ensuring that the product will meet the needs of its intended market, and after carrying out world-wide market research to locate reliable suppliers of the most cost-efficient parts, bicycle production may be undertaken at much less cost than what is required to import finished bicycles, and there will be a larger market for these less-expensive bicycles than for imported finished bicycles. As with watches, technology in bicycles in Japan started with repairs and the production of replacement parts. Those engaged in bicycle repair (and production) were former blacksmiths, lathe operators, pump makers, and makers of Japanese watches. Miyata Eisuke, a late-Meiji-period gunsmith, was so skilled he was able to make all parts except tires, rims, spokes, and ball-bearings. But he later abandoned complete production in favour of subcontracting to parts makers for reasons of profitability—not unlike the button merchants who organized separate manufacturers to carry out the various microprocesses composing button making.

The Transformation of Technology in the Process of Industrialization

Industrialization, especially the process from technology transfer to self-reliance, should be a process accompanied by transformation. This may be called the adolescent period of technology transplantation.

The degree of technology transformation ranges from a single machine in operation to the change of an entire system. One of the most rudimentary examples can be found in the first British spinning machine Japan imported. The machine had been designed to fit the height of British workers; consequently, because Japanese workers were shorter, they could use it only by looking upward and stretching their arms, which tired them quickly and diminished their efficiency. The mere installation of a footstool solved the problem, reducing worker fatigue and increasing productivity. Such a minor modification can sometimes greatly enhance efficiency and induce extremely important results.

In the effort to hit upon the most appropriate equipment for silk reeling, first an Italian big-frame, direct-reel type reeling machine was introduced at a factory in Tsukiji, Tokyo; this type required 60 workers to operate. A decision was then made to introduce at the Tomioka factory a French-type reel machine with a small frame and a rereeling method requiring 25 workers on one unit. Ultimately, all French-type joint spinning and reeling equipment was replaced with the Italian kennel-type machinery. As Okumura Shoji, a technological historian, states:

A thorough, comparative study of each owner-country's technology should have been made before introducing the machines, but it was impossible to do so in the short time since Japan had opened its doors. As a result, the Italian type, French type, and their imitations spread, and it took approximately 20 years to standardize with the most appropriate one.[48]

Technological transformation in Japan took one of the following two paths: (1) modernizing traditional technology, (2) making modern technology traditional.

With the second, the production line was subdivided into individual processes, and workers were instructed to move from one to another to become skilled in each major production process and, thus, to master the overall technology. These major production processes were then divided into smaller ones and separated from the production line to become subcontracted work. Defining the subcontracted portion as type 2a and the non-subcontracted as type 2b, much like the relation between cotton spinning and weaving, what is described is a system under which the former process is undertaken by a factory equipped with modern machines and the latter process by a cottage industry using, in this example, traditional hand-weaving machines. Because the functions were shared, interdependent, supplementary relations were established, and both could develop.

The first path, modernizing traditional technology, involves type 2b. A change in the energy source from human power to water power and from water power to electricity is an example. In the ceramic industry, the energy source was changed from charcoal to coal, and from coal to electricity (or, all three were used separately for each relevant process).

In the transformation of technology in Japan, there was competition for market share among the bases of traditional technologies (such bases depending on the local supply of basic resources). With modernization, the market was no longer limited to the domestic market. Here we see the transformation of technology to meet the existing market conditions. We also see here that it is not mandatory to adopt the most scientifically advanced technology in order to profit in terms of development. Indeed, we can see the necessity of an alternative technology (i.e., a modernized traditional technology or an endogenized modern technology) and how it should be concretely applied.

Regarding the problem of development, commonly, long-term policies are proposed on the basis of macroscopic analyses of industrialization, but this has not solved the fundamental question of development. The reason for this is simple: there has been no practical analysis of the initial conditions, the conditions for starting. These conditions have never been the same for all countries, all areas, all seasons, or all industries. What the broad analyses provide are only the general data regarding the results and not the conditions for beginning development.

The problem lies in the alienation between long-term theory, with its universal applicability but lack of immediate practicality, and short-term theory, which is concrete but not universally appropriate. The only solution is to gather concrete case-studies and from them attempt theorization.

12

Vocational Education and Development

Despite the efforts undertaken by developing countries to build their man-power, success often has not come as readily as expected (however, it is often said that the effect of development has been excellent in education and the military in comparison with other fields.) Consequently, of particular interest in the Japanese experience has been the question of education.

An ILO document pointed out as early as 1966 that "any success in indus-trial development depends, to a certain extent, on the possibility of utilization of skills. The productivity of a plant or facility will be low if the ability of its management or workers is low. In such a case, domestic investment will provide no assistance to development, but become a burden to a very poor society."[49]

The current situation of an abundance of unskilled workers or the incom-plete utilization of them indicates there are problems of national efforts and international co-operation in both the development of ability (education) and the manner of utilization (administration).

The solution to the manpower problems facing developing countries does not rest exclusively with the formation of industrial manpower; indeed, the formation of manpower in this way is not recommendable. Nevertheless, it is true that the formation of industrial manpower is currently an important and urgent problem.

Development and an educational system in support of it have, at times, given rise to campaigns of opposition. Not long ago in India, for example, a campaign against central-government-controlled education achieved wide-spread support.

Education is necessary for development, but when efforts to promote de-velopment and education run up against opposition, a social and national consensus cannot be formed, or any that may be formed or that already exists can be destroyed, throwing development and education into confusion.

Such problems have been experienced in Japan. There was not nation-

154

wide support for or a consensus over all facets of "modernization from above," the initiative for which was taken by the government. Often, the opposition to institutionalized education (and its quality), for example, resulted in riots.

In Tsuruga, central Honshu, for instance, followers of a certain sect of Buddhism rioted in demand of restoration of preaching by the sect, rejection of Christianity, and discontinuation of the teaching of Western-style writing. This was in 1873. In the following year, farmers in Okayama Prefecture rioted in protest of formal education, and 46 elementary schools were destroyed; the farmers were also demanding the abolition of conscription. In Tottori Prefecture, 10,000 farmers rose up against schools, conscription, and the Western calendar. Similar riots occurred in Kagawa and Fukuoka prefectures, where 34 and 29 schools were burned.

All these riots took place in the year following the initiation of the national educational system, and "opposition campaigns continued for several years after that" (Nagai 1969).

The school system proclamation of 1872—which "aims to realize, whether commoners, nobles, samurai, farmers, craftsmen, merchants, or women, education for every family in every town and every member of every family"—was a declaration of equal-opportunity and compulsory education for all. According to the plan, the whole country was divided into 8 university districts, each of which was divided into 32 middle school districts, and each of these into 210 primary school districts, with one primary school in each. The primary schools were to be divided into higher and lower, each with a 4-year curriculum.

The plan was more democratic and ambitious than the system in France of 70 years later. However, the lack of full financial backing by the central government created a heavy burden for the inhabitants of each of the various districts when the time came to set up the new schools. Under such circumstances, the hardship (in figures amounting to about 8 per cent of their average annual income) induced the sort of riots mentioned above. The government was able to take care of only two-thirds of the necessary expenses, and it halved the budget for the second fiscal year.

Consequently, seven years after the plan was initiated, the education system was reformed under the Education Order of 1879. The French aspects of the system were abandoned, and a system modeled on the US system was introduced.

The new system reflected the effects of the Freedom and People's Rights Movement, a nation-wide political movement of the early Meiji period involving former samurai and commoners whose aim was to reform the new government along western democratic lines. The centralized and conformist nature of the earlier system was changed, and it became possible to establish private schools. As a result, the number of schools increased, but the quality of education fell and school attendance dropped. In this way, "the Education Order went bankrupt as soon as it was promulgated" (Nagai 1969).

When Westernization of education in pursuit of rapid modernization was being forced to retreat, a group loyal to the Imperial Court initiated a movement aimed at reforming education by restoring the national tradition. This led to a confrontation between the Imperial Court and the government. The dispute had a decisive and long-lasting effect on the course of Japanese education.

Thus, the oft-heard belief that the Japanese success in industrialization and modernization was all thanks to the successes in education is an oversimplification. Neither education nor industrial technology has ever progressed in a straight line, and both have usually had to advance along tortuous routes.

Recognition of the success of Japanese education must not neglect the role of the military in education, especially its dissemination of the technology of machine operation and its role in teaching the concepts of hygiene and punctuality. Acknowledging its positive role does not, of course, represent a denial of its negative aspects. In addition to the military, the role of vocational and technical education, which took place not in the schoolroom but at the production site, was also critical.

Typical of this on-site educational process is the practice of rotating engineers, for example, from the research and development department to the shop-floor and back again; this practice reflects the importance attached not only to the basic science but also to the practical aspects of a technology that must be considered if operation and productivity are to prove efficient.

In the early stages of educational standardization, technical education was not included in the national education system integrated under the Ministry of Education. Each ministry had its own policy and organization to cultivate manpower for technology transfer. Vestiges of this can be seen even today in the Tokyo University of Fisheries, under the control of the Ministry of Agriculture and Fisheries, and the merchant marine colleges, under the Ministry of Communication and Transportation.

The Ministry of Education was established in 1871, which, it might be noted, followed by just a year the introduction in Great Britain of the compulsory education system on a nation-wide scale. At the time, vocational education and ordinary education were separate systems in Japan.

Despite this lack of a unified education system, however, each was successful in absorbing and disseminating new knowledge and theories. Indeed, a segment of the population had attained a certain level of education, although only a select few. The seeds had been sown; it was necessary to spread elementary education to encourage growth and maturity. Throughout this process, the principle of attaching importance to the production site in vocational education was maintained. After much time, an integrated national education system was established, one reflecting the evolving needs of national development. In the course of its establishment, there have been immense and diverse technological changes, and it has now become necessary for the entire system, from primary school to the university (including postgraduate courses), to meet the needs of the high degree of basic scientific knowledge required in the face of rapid innovation in technology and engineering. Thus, in Japan the seriousness of the issue of education is fundamentally the same as in developing countries; the only difference is in the form and circumstances of the problem.

Japanese Modernization and Education—Take-off and Fall

Needless to say, those educated at public expense should, ideally, contribute, in return, to fulfilling the needs of society in a manner and degree that befits their education. And indeed, in the early years of the Meiji period, this goal was satisfactorily realized. The foreign scholars, engineers, and specialists employed by the government and private companies at great expense were able to be replaced in a relatively short time by Japanese who had received training and education abroad. The students sent abroad felt themselves to be an élite group; they possessed a keen sense of patriotic duty and responsibility. It was fortunate that education in science and engineering had been institutionalized in the West since the middle of the previous century; the timing could not have been better. This was particularly so with regard to Germany, where a number of students sought training at some universities there. A late comer in western Europe compared with Great Britain and France, Germany at the time was entering the industrial revolution, and its engineering science—rooted in national culture—was attracting the attention of other nations. It was lucky for Japan, an even later entrant than Germany, to be able to learn from the experiences of a late comer.[50]

The technology then was simpler than today's. It consisted of combinations of skills, and once the principle of integrating them had been mastered, the only obstacle remaining was to establish links within and outside each specific technology. Japan's traditional technologies and the necessary resources were at hand, and a national consensus and government policy for development were stable. Yet, despite these favourable conditions, it took more than half a century to establish even a basic system of national technology. Nevertheless, Japan's experiences seem to show that international co-operation is important for creating an independent national technology.

The darker side of educational development at this time was the emergence of a nationalisic bent. Isolated from the rest of the world, Japan grew into an imperialistic nation. The Meiji education system was controlled by the government, which moulded it to suit its nationalistic aims at the sacrifice of the cultivation of humanity and talent. In fact, in the nature of the "splendid take-off" educational development experienced were the seeds for its "fall" (Nagai 1969). How this was so should become clearer as our discussion progresses.

Despite the government's manipulation of education, technical and vocational education continued to progress. Science and engineering, by virtue of their rational, logical nature, were protected from the "supralogical" state religion of Shinto. Perhaps for this reason, students from Korea and Taiwan, then colonies of Japan, chose to study these subjects. Many such students studied medical science because the big companies had not yet opened their doors to them, even though they had graduated with degrees in engineering and science, and it was easier to set up practice as an independent medical doctor.

The Meiji government was confronted with the problem of the revision of the unequal treaties with the Western powers. It was also being challenged by the growing activity of the Freedom and People's Rights Movement, which was demanding greater freedom to participate in national politics. After the Southwestern Rebellion of 1877, its supporters (Okuma Shigenobu and others) were expelled from the government, which evaded a crisis by promising the establishment of a national diet.

The leadership in the Freedom and People's Rights Movement shifted from the ex-samurai class to the wealthy farmer class, and a movement formed to draft an original constitution; violent uprisings arose in various places between 1881 and 1889. Peace came with the promulgation of the constitution (1889) and, in the following year, the election of Diet members, which was immediately followed by the promulgation of the Imperial Rescript on Education.

The liberating effect of the Meiji Restoration caused ex-samurai and wealthy farmers to question the grounds of the government's legitimacy. One result of the ensuing struggle was the establishment of government control of education and the institution of Shintoism as the state religion and as the principle for education. In response to an inquiry by the Emperor Meiji in 1879, prior to the promulgation of the Imperial Rescript on Education, Motoda Nagazane (1818–1891), a Confucianist, declared in his *Outline of Education* (also called *Imperial Will of Education*) that the "state bureaucracy properly holds the ultimate authority" in education. With the creation of this document came the loss of the power to check "the control by state bureaucrats" (Nagai 1969). The combination of this bureaucratic hegemony in the national education system and the politicization of Shinto became the Achilles' heel of Meiji education.

Seen from another angle, what was occurring was the politicization of bureaucracy. The national Diet was held to the mere status of responding to inquiries and was thus not a legislative body, while the bureaucrats, no longer functioning as administrators, were transformed into a political force, especially those of the Home Affairs Ministry, and began to intervene in education. The Home Affairs Ministry (established in 1873 and abolished in 1947) was the general headquarters of the police administration, which was in charge of maintaining public security and political order. Its ferocious nature peaked in 1928, and this was the time when "maintaining law and order" was used as a pretext for repression and thought control by the "special police."

After the Imperial Rescript on Education, education and most of the country's cultural institutions were made into political institutions, and the age in which "Japan eagerly exerted itself to learn from the West" had ended (Nagai 1969).

Vocational Education and the Normal School System

Although established as a part of the vocational education system, the apprentice schools were unable to fulfil their promise because they were forced to supplement the elementary schools, which were handling only about 50 per cent of the educational needs at the time. Although the schools aimed at transforming skill training under the apprenticeship system that was common to the traditional craft technologies to the cultivation of technical skills under a school system, assiduous efforts by highly talented teachers were required to banish the antipathy that "learning is not necessary for craftsmen" (Sato 1982; Toyoda 1982, 1984). Native technology and industries had to be integrated with modernization.

At the turn of the century, Japanese traditional industry started to fluctuate in parallel with the rise and fall of the modern economy, and modern industry soon began to lead the way. The apprentice school was upgraded from the level of a mere supplement to elementary education to middle-school-level vocational-technical continuation school. It was in these schools that the future leaders in local industries received their educations and training (Toyoda 1982).

As the example of the Seto Ceramics School in Seto indicates, it was due to the efforts of vocational school graduates that traditional industry was successful in adopting modern technology, renovating its management, and advancing into new products, thus transforming itself into an export industry.

It has been generally accepted among Japanese scholars that the apprentice schools and technical continuation schools "did not play an important role as a supplementary educational organization for shop workers" or "did not play any role at all."[51] However, an examination of traditional industry forces a reversal of this evaluation.

The role of the apprentice and technical continuation schools was important especially in regard to the most urgent problems of development, the diffusion of modern technology, and the modernization of traditional industry—particularly in the areas of agriculture, dyeing, commerce, and local industries.

At the beginning of industrialization, military weapons' and other large factories had to establish their own, on-site technology-training centres because they could not depend on the public schools for technology education.

As I mentioned earlier in reference to iron manufacturing, the technology accumulated in this way was usually not transferable to other enterprises or industries. While among low-skill workers there was a tendency to shift from place to place, in highly skilled workers there was a strong tendency to remain in the same place, as the skills were specialized and the pay rather high. Moreover, even with the improvement in technology training in the schools, higher-level technology training and education have continued to the present within each enterprise; this is a peculiarity of Japanese education.

The teachers in the vocational schools had an orientation and personality that distinguished them from those in the general or elementary schools. As general education became increasingly influenced by politics and the social imbalance brought on by the industrial revolution became more apparent, the economic hardship of the students began to manifest itself in the classroom. The long absences from school and malnutrition forced the teachers to concern themselves with the economic and social solutions of the problems. However, this was regarded as politically dangerous. The more concerned teachers inevitably became more political, and, as a result, were repressed. The vocational school teachers were less involved politically, and had to grope, under the circumstances, for practical solutions to the problems.

Here we are compelled to turn our attention to the normal school and consider the relation between the situation described above and what education scholars refer to as "double-track" or "dual structure" education.

Double-tracking refers to education of an élite to meet the needs of the state and administration on the one hand and commercial and vocational education to meet the requirements of the general population and private sector on the other. An example of the former was the education of technocrats and techno-scientists at Tokyo University; the training of engineers and managers at the industrial and business sites is an example of the latter.

Dual structure, regarded as the "moulder of the Japanese-type of intellectual, describes the particular educational system of the Meiji state, where the degree of coercion of nationalistic education at the elementary level became weaker and the degree of tolerance for liberalism became greater as the level of education rose to secondary education and then to the university.[52] The best example of this was in the university: for an assistant professor to be promoted to the rank of professor, it was essential for him to study abroad, in Europe, or sometimes in China, at the expense of the Ministry of Education. This situation continued until World War II.

Simply put, the desire to throw off the yoke of the unequal treaties and gain an equal footing with the Western powers collided head-on with the determination to forge national unity through the application of state power. These contradictory tendencies were present in the early days of the Meiji regime, but the Rescript gave formal expression to them, and the Ministry of Home Affairs became their voice, while the ministries related to industrial and technological affairs were pushed to the sidelines. Later, the military superseded Home Affairs in the control of education because of corruption by political parties and the pressures of international relations. What had given birth to such self-righteous militarism was the Meiji education system.

The normal school in Japan was developed by Mori Arinori (1847–1889), the first minister of education. Mori, who was so eager as to be almost extreme in carrying out educational reform and development, intended with the normal school to correct the intellectual polarization and imbalance in national education. Consequently, Mori set up his Mori Arinori's Commercial Training School. And yet, despite his pioneering efforts in vocational education, he put the normal school, established to educate teachers to educate the manpower necessary to build a new nation, under military-style discipline.

This might have been effective, but Mori militarized the normal school in a strange manner. He introduced a network of "secret advisers" and selected one or more "advisers" from each class to report confidentially to the school-master on the daily conduct of fellow students, both in class and in the boarding-house. This appears to be a gloomy contradiction between Mori's ideals and actual practice, but for one involved in the reality of the chaotic society of that day, it might have seemed an inevitable policy.

At the time, freedom was not understood to mean political freedom or the freedom of thought backed by high morals and self-responsibility. It was equated with anarchical self-assertion, egotism exempt from all responsibility. We can see in this a conflict between the "pre-modern" social character of the classes of farmers, craftsmen, and merchants, who had never had a well-developed sense of public duty, apart from family and village, and the "pre-modern" sense of responsibility of the samurai élite, represented by Mori.

The claim is that most commoners who entered the normal schools were farmers, but this is debatable. From a sociological viewpoint, the peculiar manner of personality formation in the normal school does not seem to have been one that would relate to farmers (especially the landowner-farmer class) but more closely to the ruined samurai class. Indeed, it is my opinion that the low-ranking samurai—thrown into extremely poor conditions as a result of their resistance to the new government—maintained their traditional code of loyalty, changing only the object from the old shogunate to the new government.

Because power under the Tokugawa shogunate had long been consolidated, the samurai was no longer a warrior but an administrative bureaucrat, and his adaptability to altered circumstances was quick. And yet, the unstable status of teachers and their surprisingly low salaries, combined with the ex-samurais' sense of social superiority, caused them to feel resentment.

The rank of non-commissioned officer in the Japanese army, the epitome of inhumanity and impersonality, had something in common with the frustrated and grotesque personality formation among the ruined samurai at the Meiji normal school. In the military, soldiers who were graduates of middle or higher schools were brutalized in the name of the Emperor by the less-educated non-commissioned officers.

Similarly, in the schools, pupils from uninfluential and unpropertied families were mistreated by teachers educated at normal schools, while the students of rich or influential families wre treated well in an attempt to gain favour and effect improvement in the teachers' own positions.

The common element in the mentality of these lower-level bureaucrats was their conciliatory approach to subordinate personnel or pupils who displayed any disobedience or resistance. The reason for this was that the presence of this sort of defiance was enough to jeopardize the position of the teachers of these trouble-makers: any advantage the teachers may have had in the competition for loyalty could be lost. They tried to conceal the existence of any unique or unusual elements in their surroundings.

Late-Comer Investment in Education

As Hamao Arata (1849–1925), an enlightened bureaucrat, said in his address at the inauguration ceremony of the Tokyo Worker Training School in 1881: "In this country. . . our policy is not to establish factories and then set up technical schools to supply them, but to establish technical schools and send their graduates out to set up the factories." This is a clear example of the determination to give priority to the development of human resources at the initial stage of industrialization.

The first graduates from the Tokyo Worker Training School numbered 24, among which, however, only 3 could find jobs after graduation.

Obviously, the demand for science and engineering expertise was low. Similarly, there were few seeking this sort of training. Take the case of Niijima Yuzuru, who had planned to establish a science and engineering institute, but had to abandon his plan because of insufficient applicants. Because of the recently promulgated official medical licencing system, medical and pharmaceutical schools were the few exceptions that had applicants, though the number was reportedly small.

After World War I, however, the need for science and engineering departments had grown to the extent that they began to be established in the private schools. This was at a time when Japanese technology had achieved a primary stage in self-reliance, and the demand for science and engineering experts was consequently much greater. The national universities had taken the lead in this simply because they could easily meet the small need that had existed until this time.[53]

The advocacy for development of human resources by investment in higher education was made also by Sano Tsunetame (1822–1902), a politician, and Yamao Yozo (1837–1917), an enlightened engineer. They were themselves the product of human resource development policies undertaken by feudal clans at the end of the Tokugawa shogunate. They advocated an expansion— beyond the old political boundaries of feudal clans—to a national scale of the development of human resources. The Meiji government's development policy, the "encouragement of industry," was thus not an idea that originated with the new government, but an enlargement on a nation-wide scale and a centralized development of the earlier experiences in the former feudal clans.

However, investment in higher education was fragmented and unsystematic at first. Each ministry had, under its supervision, its own institution for recruiting and developing needed manpower. By 1878, the tenth year of Meiji, these schools (the "seven peaks competing for the development of manpower")[54] under the direct control of the various ministries consisted of: the School of Engineering under the Ministry of Works; the University of Tokyo (a unification of the former Kaisei School and the Medical College) under the Ministry of Education; the School of Law under the Ministry of Justice; the Sapporo School of Agriculture under the Ministry of Development of Hokkaido; the Komaba School of Agriculture under the Ministry of Home Affairs; the Military Academy; and the Naval Academy.

In 1886, the imperial universities were established, and the University of Tokyo absorbed the School of Engineering and Komaba School of Agriculture (the Sapporo School of Agriculture came under the control of the Ministry of Education in 1895). Universities both in name and reality were established. A comparison with universities in Europe, whose long traditions go back to the Middle Ages, reveals the secularist nature of Japanese universities.

Various Buddhist sects had had their own seminaries for priests since the seventh century, but they were outside the education policy of the Meiji government. And there were private professional institutions of Shintoism that engaged in activities not found in the neutral departments of religious studies in the imperial universities. This was a paradox of the Meiji educational system and a remarkable indication of the conformity (recalling that the Meiji government had made Shintoism the state religion) in most Japanese private institutions to the current political orientation.

Tokyo Imperial University became "the first university in the world, excepting the United States, that included a college of engineering" (Nagai 1982).[55] Nevertheless, its emphasis was on building up its faculty of law, which was the nursery for bureaucrats in the state administration.

Its graduates could be referred to as "social engineers"; they were modernizers, aiming to upgrade society based on a new, Western-oriented value system, rather than intellectuals devoted to academic inquiry and the creation of new values. However, in the first half of the Meiji era, jurisprudence had an enlightened nature different from that existing today in Japan; it supplied men through whom the rule of law and the maintenance of order by law became actualized on a broad scale, and there was a great concern with international law. During the Russo-Japanese War, scholars of international law were posted at the headquarters in the forefront of battle. Although this was nothing but a precautionary measure that a small late-comer nation had to take under the restrictions of the unequal treaties, it is noteworthy that there was such a monitoring system. It was after Japan won the Sino-Japanese and the Russo-Japanese wars and revised the unequal treaties that such checks stopped functioning.

It is noteworthy that the imperial universities had agriculture departments besides their departments of medical science, science, and engineering. This reveals the pragmatic nature of universities of those days.

Apart from the imperial universities, and even before their establishment, it was apparent to policy makers and the few techno-scientists who were in existence that there was a national need for engineers. The Tokyo Worker Training School was at the forefront in the efforts to accumulate international science and technology information, draw up science and engineering policies, and create a fund of trained personnel. It developed manpower for manufacturing and trained engineers who could supply the needed technological support and leadership.

The Tokyo Worker Training School (which became the Tokyo Institute of Technology) was established in 1881, offering a one-year preparatory course and a three-year regular course, as an institution of technology training for developing future vocational school teachers, plant foremen, and floor supervisors.

This school became the model of practical education and the prototype for the elementary and higher vocational training institutions established throughout Japan. It also became the supply source of teachers for those schools. And, because of the great shortage of qualified teachers, graduates from the school were able, through the alumni association, to teach in a variety of places, and, as each move meant a promotion, they also won a steady rise in social status.

The traditional label *shokunin* (craftsman) yielded to the more current-sounding *shokko* (workman). More than "factory worker" it referred to the artisan or manager armed with new technology, the educated professional of modern technology.

The Tokyo Worker Training School had an apprentice school under it and was later transformed into first a technical school, then a higher technical school, and finally the present institute of technology. It came to be referred to as the MIT of Japan. Even bureaucrats such as Inoue Kowashi (1834–1895), who was eager for a vocational education system for industry, and though there was no shortage of funds once the necessary legislation was in place, could not have achieved this had it not been for the school's enterprising leader, Tejima Seiichi (1849–1918), who regarded industrial education as his mission.

Tejima established himself as a great educator because he "enlarged the concept of technical schooling to mean industrial education" (Miyoshi 1983). Without his personality and philosophy, the founding of the institute would not have materialized.

However, in the beginning, education in industrial technology was not the main part of vocational training. The agricultural school had the largest number of pupils, followed by the commercial school. Each had developed in connection with traditional local industry. As for the overall structure, there were high schools of agriculture and forestry, of commerce and of industry, and these were then organized into colleges that specialized in each field of technology. This was in parallel with the system of middle school, high school, and imperial universities. These two systems were completed between the 1910s and 1920s, exactly the time when the national formation of a technology network, the first stage of self-reliance in technology, was accomplished.

It should also be noted that Tejima and others set the starting point for vocational education at the level of elementary school. Tejima successfully introduced manual arts and handicraft training into elementary school curriculums. He understood that skill training was an indispensable part of successful technology development and transfer. Because of the long time required to develop skills, it is necessary and important that their cultivation be started at the elementary school level. This conviction on the part of Tejima and his colleagues was in contrast to the situation in Great Britain, where vocational education was merely one measure to relieve the poor and the relation between technology and skill was overlooked.

Because Japan is a monolingual society, questions of terminology and language do not occur regarding science, mathematics, drafting, and manual arts education in the schools. In a multilingual society, however, there may be difficulties and much debate over the terminology of science and technology (and the language of instruction), not only in regards to the elementary school level, if such education begins there, but even at the high school level too. We have pointed out the importance of native engineers to a nation's technological self-reliance, and we are fearful that education in science and technology in a foreign language might create an obstacle. Using foreign-language texts at the initial stage of education in technology might be unavoidable, but it is not wise to impose the mastery of a foreign language on engineering students at every stage.

"Of the major tasks in education—textbooks and methods of training teachers, for example—undertaken in Japan at the beginning of the Meiji period, nothing was worked out or designed by the Japanese themselves" (Nagai 1982). This was the reason the attendance rates did not reach 50 per cent for more than 20 years. Even the text used in the Japanese language classes was a translation of the Wilson reader from the United States. Under such circumstances, when education does not relate to a people's daily life and culture, it will not be viable. Many excellent translations and adaptations of children's stories, folk songs, and similar works could be mistaken, even today, for original Japanese works; but these examples represent exceptional cases, and demonstrate that only those adaptable to the everyday life of the people and to the national culture will survive. Japan endeavoured to cull such works for more than 20 years.

Although not directly connected with education in science and technology, the compilation of a national language dictionary was a great undertaking in Japan from the standpoint of national education. In the twenty-fourth year of the establishment of the new Meiji government, Otsuki Fumihiko (1847–1928) completed the first modern Japanese language dictionary, the *Genkai*.[56]

Regarding the diffusion and development of education, Nagai (1982) states that "when power from outside and above was added to change from inside and below, Japanese education was modernized for the first time, which was characteristic of an underdeveloped country."

Our research on development and education has stressed "change from below" and regional effects. What Nagai calls the modernization of education we interpret as the "popularization of education." In fact, modernization should be the same as popularization. After the bitter experience of "ultra-nationalism in education," the Japanese began to believe that the modernization of education should be an activity for national development based on the firm principles of democracy and peace, but with the flexibility to adjust to the changing international setting. In this sense, Japan has not yet completed its modernization of education; the state continues to control education for use in serving its own needs.

On-Site Training

At the initial stage of industrialization, when there was a small group of techno-scientists and engineers on the one hand and a great many unskilled and unemployed workers on the other, some of the skilled technicians took on apprentices and thus attempted to bridge this gap. The apprentice system was seen, therefore, not only in traditional industry but also in big, modern factories.

However, these masters and their apprentices were unfamiliar with completely new technology and skill areas, and skill formation required education and training. To meet this requirement, training centres were established within each enterprise or factory.

At the apprentice school and the technical continuation school, training was separated into study and practical application (no such distinction had existed in the apprenticeship system), and instruction was by engineering experts. There was a significant difference between skilled labour trained through apprenticeship at the time when elementary education was not widespread and those who were trained in the factory training centres or in middle schools. The education effects were evident in the latter group, whose workers had gained an awareness of the logic of technology hidden in the empirical reality.

The basic training for Japan's special brand of engineer was received in the schools established within each enterprise—equivalent to secondary education—and through the vocational education system. This training was far from complete, but the long process of skill formation got its start there. Not only the leading engineers of big enterprises but also the craftsmen who pioneered the technological progress in the traditional industries were trained under this vocational education system.

When the inner and outer links of technology had been established on a nation-wide scale, these vocational schools were upgraded to colleges. For students who had worked their way up through vocational school, technical school, and then college, the college course work consisted of mainly subjects not directly related to their technical specializations, such as a second foreign language, and, as they had by this time already gained a firm grounding in the core technical fields, they could turn their attention to these other subjects. In other words, the college started supplying "educated professionals." Besides bankers, for example, the Tokyo University of Commerce also graduated diplomats.

In 1908, Tokyo Imperial University began offering an economics course in its faculty of law; it introduced a commerce course in 1909, and newly established the faculty of economics in 1919. Here, professional education from above and from below intersected. In 1920, the Yokohama Technical College (currently the Faculty of Engineering of Yokohama National University) was established and a unique system of no examinations and no marking adopted. Contrary to what one might expect, the system did not encourage idleness and the students received high evaluations. It was a symbolic event in the most liberal age of education.

The question of education in Japan was very often the subject of our discussions, and many cases in which Japan had been misunderstood were rooted in an overestimation of the education system and national policy toward education. In truth, the most important elements were the shop-floor training in the formation of skills and the training of engineers on the basis of general education.

13

The Development of Japanese-style Management

The Japanese Approach

For some time now, the characteristics of Japanese-style management have been a popular topic, mainly in Europe and in the United States. Such topics as the business group, the seniority wage system, the lifetime employment system, the periodic recruitment of new graduates, and the *ringi* system have been examined in diverse ways. But a look at the actual operations of enterprises in Europe and the United States indicates that, though there are differences due to managerial climate and environment, there are many similarities of principle between the Japanese and Western approaches. There is even the opinion that it is inappropriate to conclude that Japanese management practices are special or that they are not universal. Emphasis in recent analysis has been on how the Japanese management style has arisen and evolved historically, rather than on its typological characteristics.

Economic Nationalism

The problem of Japanese management methods, in contrast to those in Europe and the United States, and the formation and development of a government-led national economy have become an issue of concern, especially when examined in the context of the situation in developing countries. And yet, Japanese bureaucrats complain about their lack of power. A Japanese economic bureaucrat lamented, after making a visit to the Republic of Korea, that what Korea is doing would be impossible in Japan.

If the government's policies encourage development and if industry and management respond co-operatively, the relation is said to be good. The comment on Korea, however, implies that it is now not good. Indeed, one sector of Japanese bureaucracy disparagingly compares the present business world to the military of the past.

168

Since the Meiji period, Japan's development as a nation-state has been supported largely by government enterprises. Private enterprise has wielded little power, and an atmosphere of predominance of state power over the private world has prevailed.

The bureaucracy has been criticized as lacking flexibility; it has no objective standard for evaluation, and tends toward expansion. Although it professes a concern for public welfare and proclaims its neutrality, it cannot escape a conservatism rooted in legalism and bureaucratism. It is difficult to expect bureaucrats to respond quickly to changing situations. Criticism from the business world triggers such comments from the bureaucracy as the one on Korea. In other words, the physiology of the bureaucratic system does not respond well to economic logic. Unlike in the United States and elsewhere, Japan has no tradition of representatives from specific industrial fields entering the administration to take charge of policies. In Japan, the bureaucracy is made up of professional bureaucrats.

Nevertheless, foreigners opine, Japan is something best represented by the label "Japan Incorporated."

Although the make-up and logic of government and business are basically different, the two once enjoyed good relations in Japan. But these relations have been lost, and the government's leadership has been failing.

In the Meiji period, the government adopted preferential measures to assist the Mitsubishi Company in opening up coastal navigation. Mitsubishi actively co-operated in order to bring about realization of the government's (Okubo cabinet's) policy of guiding and encouraging small enterprises to devote themselves to industry.

The government undertook to sell certain state enterprises to the private sector, and companies that qualified followed Mitsubishi's example. These companies had managerial ability and were ready and willing to co-operate in the national programme.

The consensus among politicians and entrepreneurs in this period was to curtail imports and promote exports. Thus, in terms of policy, the national interest did not conflict with the interests of private business. One attraction for the private sector was that the government had collected technology and information on the international environment and economy in an era when information was scarce and valuable.

The Impotence of the Political Parties

If the shareholders of Japan Inc. were the Japanese people and the party politicians were the agents attending the general meeting of the corporation on behalf of the people, the functions to ensure long-term dividends would be performed by the politicians and their parties. However, both before and since World War II, the Diet has not attracted the intellectual élite, and the political parties and politicians have been incompetent in planning policies.

Consequently, the bureaucracy at the central government offices has been engaged in policy planning on behalf of the politicians. This has created the inclination and possibility for high-level bureaucrats to become members of the Diet. Once the major economic and labour organizations (e.g., Keidanren, Nikkeiren, Sohyo, etc.) were able to collect information, the economic bureaucracy started performing the functions of long-term forecasting, planning, and co-ordinating.

Nevertheless, the central government bureaucracy has been able to wield important influence only twice during its recent history: once in the mid-Meiji period and again after the Second World War, when the bureaucracy was relied upon to carry out the administrative tasks relating to Allied policy.

The partnership of government and business has two aspects, one fixed (the independent, divergent tendencies of each), the other (their common interests) changing as time passes, and insofar as the protection of national interests is concerned, the two have been compatible. However, when excessive national interest drove the entire country to a wartime economy, a group of bureaucrats who thought of themselves as "new bureaucrats" sided with the military. This was a time when enterprises, while protected, faced extreme limitations of activity under the controlled economy. At about this same time, the *zaibatsu* were beginning to be forced to loosen their firm economic grip.

The formation of a wartime economic system began in 1931. In connection with its construction, the *zaibatsu* came under fire. There were radicals in the military who were highly critical of the monopoly of wealth and economic control by the *zaibatsu*; the younger officers, in particular, felt that, to increase revenue, taxation of the high-income class must be undertaken. As a result, moreover, dividends income earned by *zaibatsu* families was targeted for high taxes.

On another front, as a result of the government's policy of promoting the heavy and chemical industries to strengthen the military, the *zaibatsu* concerns had to make immense new investments to maintain their positions in the economy. Under the growing demand for funds, *zaibatsu* families and *zaibatsu* holding companies were compelled, for the first time, to make shares available to outside, but related, companies to raise funds. Thus, the investment monopoly in *zaibatsu* holding companies held by *zaibatsu* families was broken (Yasuoka 1981).

Although the *zaibatsu* were dissolved by order of the Occupation forces as part of the democratization of the economy after World War II (1947), a decade later, they had begun to reform, but now no longer under the control of the old *zaibatsu* families. The age of capitalism without capitalists had arrived.

The Transformation of the *Zaibatsu*

In response to a questioner who asked about the basic difference between the *zaikai* (financial circles) of the pre- and post-war periods, Sato Kiichi (1894–1974, former president and chairman of the board of directors of Mitsui Bank, the nucleus of the Mitsui *zaibatsu*) answered with: "Have the *zaikai* existed since the war? The answer is, No."

What he meant was that the true *zaikai* were a pre-war phenomenon, when closed-door discussions between the gigantic Mitsui and Mitsubishi would find reflection in policy for the whole economic world. Each *zaibatsu* had important political connections. For example, for Mitsui there was the Seiyukai party, and for Mitsubishi the Minseito.

The rankings among the *zaibatsu* were clearly reflected in the choice for the position of chairman of the board of directors of the Industry Club of Japan, one of the *zaikai*'s most important bodies (Sakaguchi 1976).

The establishment of this club, in 1917, symbolized the firm position the manufacturing sector, including mining, had attained in the Japanese economy. Before that, the economic world had been dominated by bankers and the Tokyo Chamber of Commerce.[57] The first chairman of the Industry Club of Japan was Dan Takuma (1858–1932), of Mitsui. Dan was an engineer and a leader in the modernization of the Mitsui coal-mines; he later became chairman of Mitsui and Co. As representative of the *zaibatsu*, he occupied the highest position in the club for 15 years, until his assassination by a rightist. Kimura Kusuyata (1865–1935), of Mistubishi, succeeded him.

Before World War II, the *zaikai* formed a society closed to non-*zaibatsu* enterprises.[58] Miyajima Seijiro (1897–1963), president and later chairman of the board of directors of Nisshin Spinning Co., was one of the few outsiders who unflinchingly protested against the *zaibatsu*'s despotism.

Since his university days, Miyajima had been a good friend of Yoshida Shigeru, Japan's first prime minister after World War II. Upon Miyajima's recommendation, Yoshida appointed Ikeda Hayato to the position of minister of finance. Ikeda later became the prime minister and the creator of the income-doubling policy that contributed to the Japanese economy's rapid growth.

Shimomura Haruo (Ministry of Finance) and Okita Saburo (Economic Planning Agency) were among the outstanding figures in the economic planning bureaucracy under the Ikeda administration. Their accomplishments parallel the many splendid contributions made by the so-called professors group of Arisawa Hiromi, Tsuru Shigeto, Nakayama Ichiro, and Tohata Seiichi in the years of the priority production system after World War II.

The four most influential organizations in the *zaikai* since the war are: the Federation of Economic Organizations (Keidanren), the Japan Chamber of Commerce and Industry (Nissho), the Japan Federation of Employers' Associations (Nikkeiren), and the Japan Committee for Economic Development (Doyukai). There were no traces of *zaibatsu* domination and exclusivity in these groups, and, in fact, the leaders in the reconstruction of the Japanese economy after World War II came from non-*zaibatsu* enterprises. As a result of the dissolution of the *zaibatsu* and the banishment of war criminals from public office, "only the younger generation of top-level enterprises and the leaders of second-ranking enterprises remained in the business world, and they renewed the management class" (Hara 1977).

When the low living standard and a desire for greater freedom intensified labour discontent in the period of post-war confusion, old-style managers did not know how to respond. A general strike on an unprecedented scale was planned for 1 February 1947. The labour leaders regarded the Occupation forces as the "liberation forces," but when these forces banned the general strike, a chill was cast over the labour movement and its expectation of easy liberation.

Immediately after the general strike plan failed, the Japan Federation of Employers' Associations was established. To prevent a revolution, the new business leadership—the "fighting Nikkeiren"—resisted the labour offensives, while, in another camp, younger managers affiliated with the Doyukai "groped for the greatest common measure by which to collaborate with labour." This gave rise to the notion that Nikkeiren and Doyukai constituted a sort of two-horse carriage, brought about by business's effort to avoid an economic crisis.

During the recovery and reconstruction of the economy, key industries established their operational bases, and newly formed business groups engaged more and more in big projects. When the Japanese economy entered its rapid growth period, Keidanren began advocating an equal relationship with the United States, Japan having extricated itself from its dependence on this Western power. The age of "big business" had come to Japan. The year was 1964.

Japanese-style Management Today

During a discussion in 1969 between Shishido Toshio, then chairman of Nikko Research Center, and Mimura Yohei, former president of Mitsubishi Corporation, when Mimura was asked whether a Japanese-style general trading company was possible in the United States, he said: "Because management in the United States tends too much toward the short term," it would not be possible.[59] Planning in Japanese companies often requires a long time from conception to implementation, so that a company of this kind may not conform to the interests of shareholders who expect high dividends quickly.

Shishido remarked, considering the essential nature of capitalism, that "the capitalist selects the manager, and unless he increases profits quickly, he will be fired. In Japan, however, capitalism has no capitalists. Nobody imagines the president of Mitsubishi Corporation has been selected by its shareholders. Many companies speak of their social responsibilities, but not of the responsibilities to their shareholders. Personally, I don't think this is a good thing; on the other hand, one could say precisely because this is the way it is in Japan, Japanese companies have achieved prosperity and increases in productivity" (Shishido 1970: 194–95).

Another insight into the Japanese style of management is provided by an incident involving an employee of a commercial firm who, suspected of graft, committed suicide and left a note saying the company was "immortal."

A company president indifferent to shareholders and a sacrifice of one's life for the benefit and prestige of one's employer seem especially indicative of a particular Japanese management culture.

In Japan, the number of shares held by individual stockholders is extremely small, and for each listed company, an average of 10 institutional share-holders, mainly banking institutions, hold the overwhelming majority of its stocks. It is common for companies to hold each other's stocks. For this reason, Japanese capitalism is often referred to as "trust capitalism" or "no-capitalist" management capitalism.

A strong sense of belonging, what might be called enterprise familism, and an intense loyalty within an organization that operates according to quasi-family principles and practices could account for the case of the employee who committed suicide. And he was ashamed of himself for having been suspected of a "lack of virtue." He also adhered to the code of bushido, by which he intended, through his suicide, to morally prosecute those who suspected him. Or, it may have been that his sense of responsibility toward the company caused him to want to protect the company's integrity at the cost of his own life. Bushido mandated a manifestation of one's loyalty.

In Japanese culture, a person may not speak ill of the dead, and so if a verification of the facts reveals a dishonour, its disclosure would be embarrass-ing. Perhaps the person committed suicide simply because he was tired of living. If so, suicide is an allurement toward which anyone living in a highly industrialized society may tend to be tempted, and it is a disease inherent in modern civilization.

Foreigners (especially Westerners) often find it difficult to understand the psychology and logic of Japanese group formation. That there is competition between groups comes as no surprise, but many are not aware that fierce competition often occurs within them. This sort of unawareness can lead to a misinterpretation of Japanese business management. The impression we Japanese have gained, for example, from textbooks and the popular media of management practices in Europe and the United States often does not coin-cide with the actual situation.

Japanese management is characterized by lifetime employment, a seniority wage system, vague job classifications (which means an unspecified range of responsibilities and power) and groupism. It is generally true that workers select their employers, not their occupations. And this corresponds well with the practice of regular recruitment of new graduates and the training of new employees in the particular business practices within each enterprise. Be-cause the system and individual jobs in one enterprise are incompatible with those of others, there is a tendency created in employees to settle in one company, which justifies the immense educational investment made by the enterprises.

The system of seniority wages was originally based on a great value placed on experience and skills and on the assumption that living expenses would be greater for more senior employees, and it became firmly established and widespread in the period of sharp inflation. The lifetime employment system was established in 1910–1920, when the labour movement was active, to se-cure and pacify a skilled labour force. In parallel with this, unskilled, outside contract workers and temporary workers were recruited, and skilled workers of the key production sectors were deployed from the parent company to its affiliates. This inevitably led to the formation of a dual employment structure.

As in the past, technological innovation today is changing the Japanese style of management in various ways. The focus of education and training within the company is shifting from the newly recruited to the middle- and higher-level strata of employees to ensure their adaptation to new technology. The seniority wage system has been combined with a system of wages based on job function, which itself is undergoing revisions amid rapidly progressing technological innovations.

Management concepts and practice aim at a continually expanding wealth and growth potential. Technological innovation, however, brings with it fewer job opportunities in the manufacturing sector (because of the mechanization of the production process), and it reduces the office work-force, while it encourages an expansion of the planning, R. & D., and sales divisions. This is referred to as the development of a "software economy." This has been reducing direct employment and diversifying, increasing, and shortening employment periods for indirect employment. For this reason, it is difficult to forecast in which areas the traditional Japanese style of management will continue or will have to change. One certainty is that, although the priority of where one works rather than what one does will not altogether disappear, the giving of priority to occupational preference will grow stronger.

As indirect employment becomes shorter and more irregular, it demands higher wages than direct employment. If necessary, an enterprise will convert indirect employment to direct employment, thus realizing a long-term reduction of wages. This is one aspect of developing software economies, and a possible outcome of this will be the diversification and professionalization of the technical occupations, which in turn will increase social mobility.

History of Japanese-style Management

Yasuoka Shigeaki, an economic historian, isolates the following three areas of industry where the Japanese style of management was first established following the changes brought about by the Meiji Restoration.
1. In areas that held an advantage in the beginning. Products such as silk and tea, in other words, many agricultural products became goods for export. (Coal and copper may be included in this group.)
2. Industries and products that had not been affected by the international economy: foodstuffs such as salt, soya-bean paste, soy sauce, and sake; fuels such as coal and charcoal; materials for housing such as straw floor mats and wood; and native clothing.
3. Areas and their goods that were at a disadvantage in the beginning, for example, cotton and wool, which suffered strong external competition (Yasuoka 1981).

The process was most apparent in this last group, where, in the transition to import substitution and expansion of exports, Japanese-style management became firmly institutionalized.

Yasuoka focuses attention on (1) the role division between investors and management, (2) organizations for engineers, office workers, and plant workers, (3) the employment period, and (4) the wage system. An examination of each item will make clear that there were in existence domestic conditions favourable to converting an internationally disadvantageous position to an advantageous one.

The role division (1) was characteristic of big merchants, where the owners were busy with their social functions and the actual business was performed by clerks, called *banto*. The *banto* had been trained in business skills since childhood and screened from among many employees. Just as investors in an unlimited or family-owned business must bear full responsibility as a business partner, the *banto* bore full responsibility for business activities in the owner's firm.

To maintain or develop their properties and businesses, merchants commonly had their eldest daughters marry *banto*, then let these sons-in-law succeed to the property and enterprise. This was in contrast to the practice of primogeniture in the samurai class.

Only two to three per cent of all employees were able to rise to top management positions at the big merchant houses as the job performance demands were extremely tough. There was no guarantee of long-term employment; only able men were hired for life, and the time of contract renewal was also often an opportunity for discharge. The strictness and severity of the evaluation systems used at the time, as seen in the Mitsui family, for example, were very like those in force today at the large firms (Chimoto 1982).

The rules pertained to every aspect of work and daily life. Only two holidays a year were allowed. Although food and clothing were supplied, wages were extremely low. But with patience and hard work, the apprentices accepted this as the necessary period of schooling in which to master all the needed management skills, starting with working an abacus and bookkeeping, before proceeding to transactions and contracts.

Those who could not tolerate the initial training were regarded as failures. But once completing the initial training, the new managers were allowed to live outside the shops, and were provided with opportunities to establish their own families and even the right to set up their own shops.

The time frame for skill training extended to the time of the physical examination for conscription, at age 20, by which time basic occupational training was expected to have been completed in the merchant and craft worlds. The establishment of the elementary school system, however, meant that, for some vocations, the training—which began upon graduation from elementary school—was interrupted by military service. Having formed part of the social change, the new technology, which was introduced selectively, filled the gap caused by this interruption (Yasuoka 1981).

As an example of the selectivity applied in introducing technology, in the silk-weaving area of Nishijin, in Kyoto, the machines that could not produce as fine a weave as traditional machines were rejected. A mixture of new and old, internal and external technologies was dispersed throughout the production process; or, as dictated by the particular markets and characteristics of certain products, one type selected and applied exclusively (Iwashita 1982). Modern technology and machinery were carefully evaluated from management's viewpoint before being adopted; this selectiveness implies the existence of a large stock of managerial ability.

Japanese-style Management and Managers of *Zaibatsu*

The question arises, how did the stock of management ability among the old leading merchants respond to the rapid political and social change precipitated by the Meiji Restoration.

The transformation from privileged merchants (*seisho*) to *zaibatsu* as an example of this process was characterized by three conditions: (1) management based on familism, (2) family-owned and -operated enterprise, and (3) maintenance of broad, family-operated network of assets and business connections.

Management based on familism had an essential internal logic in response to the establishment of a modern private ownership system by the Meiji government, which, in order to gain credit with foreign countries, had to establish modern civil and commercial codes. Thanks to the *banto* system, for example, the top merchants had their "management specialists," able men in the right places.

On the other hand, the merchant families maintained their familistic structure, for without limitations on the execution of modern private law and private rights that allow the free disposition of properties, the preservation and development of family enterprises could not be ensured. To prevent the dispersion of family properties, these merchant houses instituted a "family constitution," or "family precepts," shunning outside capital investments in their family businesses.

Yet, under these circumstances, raising funds for expansion and diversification in response to social change was difficult, and businesses were compelled to venture into risky areas of high profitability and accumulation. The Mitsui family, aided by its able managers, was successful; the Konoike and some other banking families were not, however, as blessed with able management specialists. They could not adjust to the changing situation and were anxious only to preserve a safe and steady, conservative business management, which led to diminishing capital and finally the loss of their influence in industry.

A family's monopoly of capital investment, that is, closed management, was one type of response merchant families made to the rapid changes, a response they did not consider unusual. It was common for an owner to try managing the family business with his own capital and to prevent others from interfering with his control. Unlike other business groups, both the large *zaibatsu* and the small local *zaibatsu* were fortunate in being able to practise this independence (Seoka 1982; Fujita 1981).

The *zaibatsu* formed in the early 1900s. The relationship between the head of a *zaibatsu*, the employer, and the management specialists was not one based on the practice of modern contracts. It was closer to the relationship normally found between a master and his servants. The confusion or contradictoriness of the *zaibatsu* family members managing their managers arose from the fact that, while the *zaibatsu* families were the sole stockholders and carried full responsibility for their businesses, they left actual operation wholly in the hands of their managers, who had no ownership in the businesses whatever. Some experts (e.g., Noda Nobuo) maintain that if there had been no interference in operations by the families, the *zaibatsu* businesses would have been more active.

Regarding their legal status, the *zaibatsu* families assumed the form of an unlimited partnership, because, as such, it was not necessary to disclose the company's financial status (Mitsubishi was a limited partnership) (Yasuoka 1981). Scholars speculate variously that the *zaibatsu* families adopted this legal form to avoid having their income from property be regarded as unearned income, as a measure against taxation, or to avoid donations. The Japanese *zaibatsu* were not alone in keeping their financial records confidential. All organizations of this kind did. The Rothschilds, for example, did not disclose a balance sheet before World War II.

Zaibatsu Managers and the Reference Group

The *zaibatsu* managers felt a personal loyalty to members of the *zaibatsu* family—not unlike that demonstrated by the company employee who committed suicide. A keen national consciousness, too, however, could be observed in the managers: Enterprising managers were motivated in the early days to contribute to the nation's development through *zaibatsu* family property. If too highly motivated, however, they were considered dangerous to the welfare of the family and consequently isolated.

Because most managers were well-educated ex-samurai or ex-bureaucrats, they were familiar with foreign countries and modern technologies, and had an enlightened, modern outlook. They were thus well qualified to participate in the commercial and industrial activities of the day. Nevertheless, just as Dan Takuma hesitated to be "a clerk for a merchant family" rather than a government bureaucrat, they also had a psychological resistance toward clerking for merchants.

They were interested in social prestige and official approval from the nation and the government for their achievements. They were active in social and cultural projects outside the *zaibatsu* businesses. This was in sharp contrast to the *zaibatsu* families.

They devoted themselves to religion and philosophy (Seoka 1982). This might have been a means for the *zaibatsu* managers to maintain a psychological equilibrium between private and state interests. This desire to maintain a balance between personal and national interests seems to indicate there was some tension in existence from the time of the formation of the *zaibatsu*.

These managers had to concern themselves also with the public image of the *zaibatsu*. Thus, an ultra-right reformer like Kita Ikki (1887–1937) would receive contributions from the *zaibatsu* to maintain a relationship with him and his followers and thus lessen the potential threat Ikki and the reformists posed to the *zaibatsu*.

There were other reference groups, some made up of Christian philosophers and others of state Shintoists. In any case, "the *zaibatsu* were not risk-taking pioneers; rather, they were good at following the pioneers to harvest their results later" (Yasuoka 1981).

Local *Zaibatsu* and New *Zaibatsu*

The local *zaibatsu* and the newly formed *zaibatsu* contrasted with the big *zaibatsu*, which adhered closely to the intentions of the central government and enlarged and diversified their activities on a nation-wide scale.

The local *zaibatsu* amassed vast wealth through diversification of their business activities, but they did not move from their local bases even after they had grown into nation-wide enterprises. In general during this period (early 1900s), the local *zaibatsu* did not adapt themselves to the line of industrialization led by the central government. Because their markets were not of a size that would have required the introduction of modern technology, their diversifications were carried out mainly in the traditional industries. Their chief areas were large-scale wholesaling and retailing, brewing, and food processing. When investments in the (local) railway, coastal transportation, warehousing, and electricity were added, most of the local *zaibatsu* constituted businesses of a size paralleling the agricultural landlords.

While hesitating to invest in modern industry, these *zaibatsu* invested mainly in banking. Accompanying development, money was in constant strong demand, and the banking business, which was stable, did not require as complicated a technology as did other industries.

Once banking was generally recognized as an important institution, holding the key to local industrial development, local notables began entering the top ranks of local banks—not for the sake of financial gain so much as for social prestige and to give the appearance of serving their communities. Thus, there was no evidence of plans to take control of the banking institutions through investment to utilize them as their own financing source.

Most local *zaibatsu* became *rentiers* in the process of the nation-wide structural changes in the economy because they could not recruit capable management personnel, and the limits of their financial capacities precluded them from entering industry. On the other hand, ambitious businessmen operating local mining or silk concerns grew into *zaibatsu* by riding the wave of industrialization and making full use of the technology their families had accumulated.

The top management of these new *zaibatsu* were men educated as engineers, scientists, etc.; they were well aware of the need for new investments and technology. The enterprises they established formed new *zaibatsu* by moving into fields where their technologies were interrelated. Some of these new *zaibatsu* included Nissan (Nihon Sangyo), Nisso (Nihon Soda), Nitchitsu (Nihon Chisso), and Riken (Rikagaku Kenkyusho).

These new *zaibatsu* had strong foundations and could undergo rapid growth. They were more daring than the old *zaibatsu* as they advanced into the heavy and chemical industries, and a characteristic common to all was that their founders were engineers.

Because the technology in these industries ranged over energy, materials, parts supply and processing, and other sectors, the new *zaibatsu*, in the process of expansion, inevitably ran up against enterprises belonging to the old *zaibatsu*, and this naturally restricted their business activities. For this reason, they went in search of opportunities in Japan's colonies overseas, such as Manchuria and Korea, where they played a leading role in the formation of the heavy and chemical industries. For these new *zaibatsu*, the younger government functionaries, the "new bureaucrats," acted as their political mentors, and they, in turn, had the support of the military. The defeat in World War II meant a loss of most of the foundation on which they had been established.

From the viewpoint of the five Ms, however, this loss was not fatal for the reconstruction of the *zaibatsu* as business groups. What was decisive, though, was their inability to raise money or secure credit, which controlled all the five Ms. In other words, it was the group's lack of supporting banks that was fatal. The ex-*zaibatsu* enterprises have revived as new groups organized around the new banks.

Of these new *zaibatsu*, Riken had started out as an institute of physical and chemical research in 1917. It described its purpose thus:

To promote the development of industry, the institute will be engaged in research in physics and chemistry and in the pure sciences and their applications. In either the agricultural industry or any other industry, an institute that does not base itself on physics and chemistry can make no steady development. Especially in Japan, which has a dense population but insufficient industrial and other materials, there can be no other way to realize the prosperity of the nation than by promoting the development of industry by means of education. Our aim is to accomplish this mission.

An examination of how Riken's goal of national development on the basis of science and technology was realized deserves careful study. It perfectly epitomizes the period of the transition from importation to self-reliance in technology.

In passing, it might be mentioned that the three Nobel laureates Yukawa Hideki, Tomonaga Shin'ichiro, and Fukui Ken'ichi had direct or indirect relations with this institute.[60]

14

Development of Japan's Financial System

Development and Finance

One major problem regarding development is finance. If a country has a source of funds, obviously the problem is solved. Otherwise, it must depend on loans, assistance, or direct investment from foreign enterprises. Depending on the terms and conditions, these funds can have vicious effects on development. This is one reason we have emphasized national sovereignty as a decisive factor in the development of economic and technological independence. Although some countries have adopted development policies that leave them vulnerable to financial subordination, others have stubbornly rejected foreign capital, technology, and management funds, leading to retarded development. We live in an age of unprecedented interdependence and so must pursue a new international economic order. Because mutual dependence should be established on the logical assumption that each nation is the holder of independent politics and an independent economy, independence is the starting point.

Economic independence is not isolation; moreover, immersion into the international economy compels a nation to contend with rapid change. A country working toward self-reliance, even where a national consensus has been formed, must face a variety of difficulties. We have examined the sorts of experiences that arise as a result in the context of various industries. Turning now to the Japanese financial system and the financial institutions—after much trial and error, sacrifice, and loss, they have managed to achieve independence in the international arena.

In early Meiji, the Japanese yen had no international creditability, and therefore little investment in Japan was made by foreign enterprises, while an overwhelming deficit made the government dependent on foreign funds.

In 1899, the thirty-second year after the establishment of the Meiji government, direct foreign investment began. This was made possible by two important political and economic events that occurred late in 1898.

One was the adoption of the gold exchange standard, using the gold obtained as reparation from the Sino-Japanese War as reserve. The second was the revision of the unequal treaties after, indeed, their 40 years of existence. (The acquisition of rights for an autonomous customs tariff had to wait another 10 years.) Foreigners living in Japan were no longer forced to live in concessions, but they lost their extraterritoriality. In his *Naichi Zakkyo-go no Nihon* (Japan after the opening of the concessions), Yokoyama anticipated and warned of the severity of "free competition."[61]

Enterprises, too, at this time lacked international creditability and could therefore not float bonds in other countries; they had to depend on direct foreign investment for the introduction of foreign capital. The first instance of this was Nippon Denki Co., Ltd., which was established in 1899 with an investment by Western Electric of the United States to manufacture telephones and switchboards. Nippon Denki was modeled after Miyoshi Denki Kojo, famous for its high-level machine-manufacturing technology. Nippon Denki eventually became NEC, which, now under complete Japanese control, has long been a leading company in the Japanese electronics industry.

Standard Petroleum made a lone investment in 1900 and started the development of oil wells. "The introduction of foreign capital begun in this way was soon being used to supplement the insufficient accumulation of capital and the low level of production technology" (see Hattori Kazuma in Arisawa Hiromi et al., eds. 1967: 141).

Also at this time, the classification of banks by function, with the Bank of Japan as the core, into commercial banks, *kangyo* banks (investment and promotion of industry), and savings banks was established. It has been said that "in no other country was the banking system consolidated in terms of legislation as it had in Japan." This is one example of the advantages of being a late comer.[62]

The financial institutions organized in the 1890s were the crystallization of ideas worked out when the Bank of Japan was founded (1882) and paralleled the secondary rise of industry and the increased number of banks after the Sino-Japanese War. The first rise of industry, which followed the period of the Matsukata financial policy in the 1880s, saw the establishment of approximately 10 private railway companies and the same number of banks.

Later, forced by the intervention of Germany, Russia, and France to abandon its dominion of the Liaotung Peninsula after the peace treaty between Japan and China in 1895, Japan determined to adjust to this loss by producing weapons domestically, which led to the construction of ironworks and brought about Japan's second surge of industrialization. This included creating an integrated system of iron and steel manufacturing, the enlargement and improvement of the national railways, the reinforcement of sea transportation and shipbuilding, enlarging the communications network, and developing Taiwan, with priority on strengthening armaments.

To attain these targets, the full-scale transfer of technology and increased imports of machines and equipment were required. But the fall of the price of silver on the international market meant a sharp increase in the prices of imported goods (which, however, was advantageous for the export industry).

Before the Sino-Japanese War, the propriety of a move to the gold standard had been discussed in government circles, and a fierce debate on how to proceed had resulted. With the sharp increase of machinery imports in the private sector after this war, the group advocating a shift to the gold standard became predominant. On the other hand, it was in the interest of the cotton-spinning industry, the only industry that had international competitive power, to adhere to the silver standard since many Asian countries—especially China— which were its biggest market, were under the silver standard.

However, when India, from which Japan was importing its raw cotton, shifted to the gold standard in 1893, and when increased imports of raw cotton precluded the cotton-spinning industry from staying with the silver standard, the time for a shift to the gold standard had arrived. The Sino-Japanese War reparations provided the point of transition. Thus was the consolidation of the financial system effected. Let us briefly examine the process of technological transfer that led to this consolidation.

Exchange Companies

The earliest banks in Japan were called "exchange companies," that is, currency exchange companies. In 1869, exchange companies were established in Tokyo, Yokohama, Niigata, Kyoto, Osaka, Kobe, Otsu, and Tsuruga to finance the promotion of domestic and foreign trade. Their business was deposits, loans, foreign exchange, and nickel silver, old gold, and silver coin transactions. The companies were given the rights to issue gold, silver, coin, and nickel silver notes.

At this time, deposits had not developed, and civil strife and political changes had brought about a state of confusion that affected currencies, and thus the amount invested in the exchange companies was smaller than the amount required as working capital. The exchange companies therefore depended on huge loans from the government. For example, the capital of the Tokyo Exchange Company was 948,500 *ryo* in 1872, of which the government had contributed a total of 332,000 *ryo*. The company's investors were exchange companies that had been authorized by the Tokugawa shogunate, such as Mitsui, Ono, and Shimada. The Tokyo Exchange Company offered financing for the production of export goods (silkworm-egg cards, raw silk, tea, and seafoods) at an interest rate of 0.15 per cent per month, which was much lower than the prevailing rates.

Despite the interest-free government financing, the authorization to issue notes, and the government's easiness about returning the money, it was not possible for the exchange companies to establish themselves and grow.

Takizawa Naoshichi and other banking experts attributed the failure to the companies' being established merely on the basis of a combination of traditional Japanese money exchange practices and European institutional features, which did not properly fit the Japanese commercial structure of the time; moreover, there were few men qualified to conduct the business. Furthermore, as a result of the rationalization of the administration, the government's trade department, which had been the leading division for the scheme, was abolished.

The new government lacked stability; it was a time of rapid political and social changes. The foreign trading firms were in an overwhelmingly advantageous position in business negotiations, protected as they were by gunboat diplomacy. Finance for leading agricultural export products had an extremely speculative nature. These were hostile circumstances for the establishment of a true banking system. Nevertheless, there were a few bankers, able to overcome the instability and risks, who survived the upheaval en route to the modern age. The successful bankers had a management strategy, abundant funds, and the ability to evaluate the situation accurately. Ono and Shimada failed because they lacked these abilities.

In 1872, only three years after their establishment, the exchange companies were forced to dissolve because they could not develop their credit systems and supply the necessary funds to stimulate industry. Nevertheless, a general awareness was created that this type of business organization as an incorporation, though incomplete, was necessary and effective. It made ordinary people more aware of business organizations, the banking business, and the operation of funds, and gave impetus to the establishment of private banks.[63] For Mitsui especially, it was this first experience of the "modern age" that led to the formation of Mitsui Bank's management philosophy.

From a National Bank to the Bank of Japan

Because the new Meiji government had inherited the foreign debts incurred by the former regime, it had to guarantee these debts, as new governments around the world had done. However, unless Japan's new government could successfully deal with the accumulation of inconvertible notes that it had issued during its revenue deficit, the resulting inflation would undermine its economic foundations. Therefore, the treatment of these notes became a political life-or-death question for Japan.

To study the problem, important figures were sent abroad. Ito Hirobumi, one of those sent overseas, found in the United States that "the national bank has the privilege of issuing notes and both functions of amortizing government notes and of financing," and he recommended adopting this system as a "measure of serving two ends."[64]

In the United States, there was a need for a policy to deal with the inflation caused by the Civil War and with the over-issue of wartime bonds (to maintain their price). The new Meiji government had the same problem. It was discovered that the US national banks could capitalize the bonds issued by the government. While the bank deposited bonds with the government and the government issued the bank notes, there arose a demand for bonds, which enabled their price to be maintained. Ito eagerly embraced this system.

Proponents of a central bank to be modeled after the Bank of England opposed Ito's plan of decentralizing the note-issuing system. The points at issue were note conversion or specie conversion and decentralized or centralized issuing. A compromise was made between the two, the government hopeful of realizing the two goals simultaneously, and a plan for a national bank system was adopted. A number of national banks were expected to be established, and a huge amount of inconvertible notes were to be collected to establish the system of specie conversion and to make the banks profitable.

Nevertheless, only four national banks were established. Why were the government's expectations disappointed? The reason that Yasuda, a famous financier, for example, was not interested in participating in establishing a national bank was due to the condition of being required to convert the notes immediately after their issue, which obviously would have greatly hindered business.

Until 1872 or 1873, there had been no difference between the gold specie and note, but because of the government's reckless issuance of notes, a premium of ¥17.8 per ¥1,000 of gold specie occurred in June 1875. With the increase of imports, an immense amount of gold specie flowed out, and for this reason, the Yokohama Second National Bank, for example, could not issue notes at all.

In 1876, because of a fear that the issuance of hereditary pension bonds would lower prices in the bond market, it became permissible to allocate pension bonds for banking capital (up to 80 per cent), and the rate of specie reserve was lowered (to 20 per cent). To summarize, the government changed its policy from one of establishing a conversion system to one of increasing the number of national banks, and thus increasing the supply of currency, which was expected to enable the government to raise funds for development.

As an immediate result, 153 banks were established, and the national banks, provided with many privileges, attained good results. For some time after that, businessman meant banker—until the 1920s, when the industrialists were added.

There is the story that, after studying the banking business in a foreign country, the son of a wealthy merchant proposed to his father that they open a bank, and the father warned that, because it was a risky business to keep others' gold and silver, a storage fee should be charged. The idea of paying interest was out of the question. This is a story from early Meiji, and represents the state of the time: there were funds (bonds), but no one knew anything about banking business practices. So the government had to begin with an introduction to banking business practices and the thinking behind them. In 1872, Allan Shand, a British banker, was invited to Japan to provide guidance in bank bookkeeping, and, as a result, British banking principles gained some currency.

But the banking system itself was American, and problems arose because of the differences between the two systems. Nevertheless, the British banking system, which had been established during the Industrial Revolution when Great Britain was the leader in international trade, was considered to be not easily transferable to Japan. Despite the problems, the mixed American-British banking system sufficed to meet the requirements of raising funds for the Meiji government's industrial promotion policy.

In spite of the government's efforts, inflation continued, and, in 1877, restrictive measures had to be imposed on the national banks. Since the gross capital amount was limited to ¥40 million and the gross amount of notes to be issued was limited to ¥34 million, the Kyoto 153rd National Bank, established at the end of 1879, was the last authorized national bank.

When the Bank of Japan was established in 1882 as a central bank, it monopolized the right to issue notes, and the decision was made to limit the period of operation for the national banks to 20 years from the time of the Bank's establishment. After that, any national bank could be reorganized as a private commercial bank, but it was compulsory for each bank to redeem all notes issued by the time of reorganization. From then on, the banks were under the pressure of rationalization.

Although the national banks had been provided privileges and protective measures, they were not active in the deposits business (because they had government deposits), and, because their loans were centred on specific enterprises (whose directors were also bank directors) that mortgaged land and buildings, the funds lacked liquidity. Further, such features of commercial banking as discount promissory notes and bills of lading were not present.

Often, the national banks could not meet fund demands, and it was a general practice for small or medium-sized banks to borrow money from a larger bank—and the larger bank from the Bank of Japan—to lend at a high rate of interest. These banks were, it could be said, merely brokerage earning banks. One banking professional commented, in fact, that the national banks had the appearance of incorporated usury.

The Bank of Japan was established by Matsukata Masayoshi. As he saw it, the establishment of finance (and the consolidation of inconvertible notes) and the idea of the establishment of a central bank were inseparable. While in charge of local administration in Kyushu immediately after the Meiji Restoration, he fully realized the damage an excessive issue of inconvertible notes could cause.

In 1878, Matsukata, then vice-minister of finance, visited Europe as the assistant general director for the international exhibition in France. He inspected the financial systems of different countries and was advised by the French minister of finance, Leon Say, of the need for a central bank system and for a convertible bank note system. He was advised not to model on the French system, which was very old and would have been difficult to transfer, but to follow the system used by the Belgian National Bank, which had been recently established and for which the experiences of several countries had been considered. Here again, Japan enjoyed the advantages of a late comer.

At the time of the establishment of the Bank of Japan, Matsukata reorganized the existing banks into a tripartite system of commercial banks, with the Bank of Japan as their centre, investment, and savings banks.

Although the Bank of Japan had begun as the core bank of the commercial banks, it became the central institution for industrial financing when, after the first modern economic recession hit Japan in 1890, it began discounting promissory notes, receiving the stocks of 15 sectors, including the railway, sea transportation, and insurance, as security.

In summary, the banking system formulated by Matsukata consisted of the following:
1. The Bank of Japan and city banks as commercial financing institutions
2. Investment banks as long-term financing institutions
3. Savings banks as savings institutions for the general public

Of the three, the investment banks were the development banks. Their purpose was "to finance the capital for the initiation of business, to promote the reclamation of agricultural land or assist in its geological improvement, and to promote such businesses as silk reeling and canal and port construction, receiving houses and land as mortgage."[65] It is noteworthy that agricultural and industrial development were given an important place in this tri-partite system of banks.

This division of banks was modelled generally on the German mortgage bank and the French movables credit and immovables bank, but in many of its details, it was original. There was a tendency to view this arrangement as being too far ahead of its time, it being based more on future prospects than on an awareness of the economy of those days.

After twists and turns, however, the arrangement was realized. Clearly stating that "the promotion of industry is the aim, and the bank is the means for it" in the Japan Kangyo Bank Law promulgated in 1896, the investment bank began its business of stimulating industrial development. The Kangyo Bank of Tokyo was the central bank, and an agricultural and industrial bank was established in each prefecture. As stated in the Bank's charter, "when more than 20 farmers or industrialists wish to borrow money on joint responsibility, insofar as the applicants are credible, the loan is to be provided without mortgage on the basis of periodic repayment within five years." Thus, the Kangyo Bank was a long-term financing institution for the encouragement of industry, with immovables as mortgage.

In newly settled Hokkaido, where capital was in extremely short supply, the Hokkaido Takushoku Bank was established. As with the agricultural and industrial banks of other prefectures, it provided long-term loans with immovables as mortgage. Also, with agricultural products, stocks, and bonds as securities, it made loans, subscribed for bonds, and handled deposits and bills of lading. Before this bank was established, the demand for funds was so great and the local financial resources so poor that interest rates were as high as 20 to 30 per cent per month. Under these circumstances, the development effects of this bank were extremely significant.

When tenant issues came to the fore after World War I, neither the Kangyo Bank nor the agricultural and industrial banks would finance agriculture and farmers because of the profit-making characteristics of these banks. This gave rise to a national federation of agro-credit and development co-operatives (Chiba 1983). The circulation of funds among members may have been an ideal form of finance by the co-operative associations, but actually, financing by these organizations generated competition with other banking institutions, which led to a change in the character of the co-operatives. Their significance, however, should not be neglected.

In addition to the banks discussed in the preceding, there are a few others in need of mention. The Industrial Bank of Japan, for example, was founded in 1900 as the "central bank for industry" and its function was to introduce foreign capital, in accordance with the policy to limit direct investment of foreign capital (Asai 1983). In addition, this bank was soon playing the leading role in reinvesting foreign capital in the Japanese colonies.

The Yokohama Specie Bank (now Bank of Tokyo) was a special bank established to (1) promote direct exports in competition with foreign trading firms, (2) recover commercial rights through exchange, and (3) stabilize the market in gold and silver coins (Saito 1982).

Finally, there were the colonial banks in Korea and Taiwan (Namigata 1981).

Generally, in late-comer capitalist countries, certain social issues will arise earlier than as experienced in the advanced countries. One response by ordinary citizens to such issues was the formation of the mutual financing associations. The activities of the mutual financing companies, an advanced form of these associations, were important (Asajima 1983). The mutual financing companies grew to a rank equal to that of a commercial bank and secured for themselves a particular niche based on specific locations and areas of industry and business.

There were also pawnshops, which were highly popular with the common people as sources of funds. In addition, the role and significance of the postal savings system in Japan's development were extraordinarily great. Modeled on the British postal savings, the Japanese system has been operating since 1875, and is said to be unique for having been established as a savings institution for ordinary citizens by the government at an early stage. What is more important, however, in relation to development policies is that savings accumulated in the system were used for investment through the Savings Department of the Ministry of Finance. The government was able, on several occasions, to overcome a financial crisis with these savings.

In 1905, the per capita amount of savings was ¥9.29 (an arithmetic average, which was equivalent to a half-month's wages for a male worker in the manufacturing industry), and depositors whose saved amount was less than ¥3 made up 70 per cent of the total. (In terms of deposits per person, this was as small as one-twelfth what it was in Great Britain.) Nevertheless, the postal savings system held 30 per cent of the savings of ordinary citizens. One might add that farmers occupied an overwhelming majority of the depositors.

Because petty deposits are costly, the commercial banks were not interested in such deposits. The reason the postal savings system, which invested in public bonds and other equities at a low interest rate, could guarantee the same interest as other banking institutions was that the expenses were borne by the government's general accounts. In 1984, 110 years after the establishment of the savings system, the balance of its deposits amounted to ¥89 trillion, and, if the postal savings system may be regarded as a banking institution, Japan's is the world's biggest. Now that the government's postal savings banks are, in terms of total holdings, on a par with the major commercial and trust banks, the opinion that the postal savings system has outlived its traditional function might be expected.

Shibuya (1981) states that the division of banks modeled on the advanced countries was not based on the distinction between banks for long-term loans and those for short-term loans. The division was made on the basis of types of industry, such as agriculture or commerce, and on location, such as home country or colonies; the commercial banks, which originally should have been short-term loan banks, have become institutional banks to supply long-term loans. Shibuya (1981) holds that a special Japanese-type structure can be observed in this.

Since World War II, although the Japanese financial system has undergone marked changes, its basic nature and structure have not changed. Now, however, faced with international financial problems and the great changes in the world situation as a whole, Japanese financial institutions and the system are being tested.

Before leaving this section, I should add that, although the Hokkaido Takushoku Bank played an important role in the development of the large island of Hokkaido, it was not included in our research. At the beginning, I considered conducting a study of the development of Hokkaido as an interdisciplinary area study. I thought that such an area study would be done better mainly by scholars and researchers living in Hokkaido. But, as it was too big a burden in terms of distance, time, and expense, the study was terminated during preliminary surveys. Now, however, I must say this was a serious omission: More than 70 per cent of the deposits now made in Hokkaido are not used there; they are invested in Honshu, the main island. It is clear that there are still problems of development in Japan.

15

General Trading Companies: Their Role in Technology Transfer and Industrialization

It may seem strange to readers from industrially advanced countries to find that general trading companies are included in the range of this project. The memory of what role they played in the Japanese experience in technology transfer, transformation, and development has faded. Perhaps this is one reason their inclusion appears curious. Nevertheless, few would deny the significance of general trading companies in the development of natural resources, imports for development, and exports of plant manufacturing systems. In fact, trading companies were the leading sector for technology transfer. They were an especially important intermediary for the importation of technology. This function has not changed, but it is easily overlooked because of the strong impression that the past 30 years of exporting technology has left.

A simple example is the cotton-spinning machine, which was in the vanguard of Japanese industrialization. The number of imported spinning machines was greatly increased in 1892, and at the time, Mitsui and Co., a general trading firm, acquired 80 per cent of its total imports from the British concern Platt Brothers. It is a remarkable example of discernment and capability for a company to select Platt Brothers, a first-rank manufacturer, in the days when Japan was under a great deal of outside pressure to abide by the needs and wishes of the Western powers.

Mitsui and Co. provided personnel with special technological knowledge and played an important role in collecting information and acting as an intermediary in the import of machines and equipment. It imported not only machines but also raw cotton from China for processing and eventual export as yarn. Later, it imported raw cotton from India and the United States to increase the competitive power of its cotton products. The company thus became known as the "organizer of industrialization."

The industrialist Kaneko Naokichi (1868–1944), aiming to establish industrially based trade, transformed a small sugar importer, Suzuki Shoten, into a trading firm that was, for a time, comparable with Mitsui and Co.[66] He played the role of "organizer of industrialization" more effectively than Mitsui. Kaneko, who possessed an eagerness to establish factories so strong that he was called "the chimney man," was a maverick, daring and successful, but responsible for some spectacular failures too, and now he attracts the interest mainly only of scholars of the history of management. Nevertheless, the companies Kaneko started have grown into the present-day Teijin, Kobe Steel, Ishikawajima Heavy Industries (IHI), Mitsui Toatsu, Mitsubishi Rayon, Nissan Chemical, Nihon Kayaku, and Dainippon Seito, well-known firms even outside Japan.[67]

The preceding discussion pertains to the Meiji and Taisho periods, the age of light industrialization. In the period of heavy and chemical industries, Mitsubishi Corporation became dominant. This company established, as direct subsidiary companies for the import of technology, Mitsubishi of Germany Corp. (1920) and Mitsubishi of France Corp. (1924). The following illustrates how important were the roles these two companies played in the importation of technology from the advanced countries and in the gathering of information.

These two subsidiary companies were crucial intermediaries between the Mitsubishi group in Japan and the various European companies. For example, by 1937, Mitsubishi Shipbuilding and Mitsubishi Heavy Industries had, between them, purchased about 30 patent rights and manufacturing licences; the arrangements and negotiations for these purchases had been undertaken by the two subsidiaries. Likewise, Mitsubishi Electric concluded a technology co-operation contract with Westinghouse, giving Mitsubishi the selling rights and allowing it to "develop the market." In the chemical industry, where the Mitsubishi group was weak, the European subsidiary companies purchased the manufacturing licences for electrolyte cells, gas generators, and calcinating furnaces from first-rate machine manufacturers. The same was true of iron manufacturing equipment, machine tools, aircraft, and weapons.

Although Mitsubishi's imported technology was exclusively intended for the formation of heavy and chemical industries for the Mitsubishi *zaibatsu*, "it provided, insofar as it did not compete with the group companies' interests, technology and information to military manufacturers and other group companies and developed a new market; in this way, it played the role of organizer for the development of the heavy and chemical industries in the Japanese economy."[68] In the process of achieving heavy and chemical industrialization on a minimum or most-appropriate scale to meet national needs, the trading firm as an organizer, as a storehouse of detailed information and economic rationalism, was indispensable. No other organization could have fulfilled this role.

The general trading companies evolved in accordance with the development of Japanese industry and technology, undergoing changes in their role, activities, and priorities, and success as well as set-backs, and this will likely not change in the future.

In sum, Mitsui and Co. was the pioneer in importing machines; Suzuki

Shoten was a courageous forerunner in plant system imports; and Mitsubishi Corp. led the heavy and chemical industries.

Since Japan entered international trade as a late comer, such related businesses as marine transportation, insurance, and foreign exchange, which had been in the hands of foreigners, were underdeveloped. Under these circumstances, the general trading companies had accepted from the beginning as self-evident that "in terms of management concepts, the national interest (industrial autonomy) should be the determining factor, and in terms of management strategy, diversification should be the governing principle" (Katsura 1977).

Though the establishment and development of links among industrial sectors is an urgent issue in the developing countries, the subject of general trading companies was, we found, among the areas that attracted the biggest concern. It is well known that in the newly industrialized countries of both Asia and Latin America, the governments have carried out measures to promote general trading firms. In the United States, even, some members of Congress have drafted legislation to promote export trading companies.

The great number of Japanese general trading companies, their large volume of trade, and the lack of similar enterprises in the rest of the world have made them conspicuous.

Recent sales-volume reports of Mitsubishi Corp. puts it at number one among Japanese companies and among the top five companies in the world, including manufacturers. In July 1980, an American economics magazine ranked it second in sales volume to Royal Dutch Shell among the 125 companies whose headquarters are outside the United States.[69] Even if US enterprises are included, only Exxon and General Motors surpass Mitsubishi.

The breakdown of Mitsubishi sales in 1980 was domestic transactions 43.6 per cent, foreign trade 56.4 per cent.[70] Foreign trade can be further broken down into imports 32.8 per cent, exports 17.3 per cent, and transactions between foreign countries 6.3 per cent. The company reported a sales volume of US$46 billion and has 63 offices in Japan (the head office, 2 major branch offices, 19 minor branch offices, 29 local offices, 3 offices of resident representatives, and 9 special project sites) and 142 offices overseas (12 branches, 69 offices of resident representatives, and 28 local enterprises and their 33 branches).

In 1980, the firm had 9,682 employees (of which 3,710 were female). Including the 2,544 employees dispatched to other domestic companies in the Mitsubishi network and the 3,099 to companies overseas, the gross number of company employees amounted to 15,325. From these figures, the per captia annual sales amount was ¥787 million (at the yen-dollar conversion rate for the same year).

Although there are annual fluctuations, Mitsubishi's biggest item is fuel (such as petroleum and LNG), which characteristically accounts for approximately 30 per cent of total transactions. Next to fuel are metals and machines, which occupy more than 20 per cent; food and animal feed, more than 10 per cent; then chemical products, textiles, and materials, less than 10 per cent. This composition of sales has continued unchanged for several years. The dependence on foreign trade has averaged around 60 per cent.

At Mitsui and Co., metals account for more than 25 per cent, petroleum and gas 20 per cent, machines less than 20 per cent, food about 15 per cent, chemical products more than 10 per cent, and textiles 4 to 5 per cent; the dependence on foreign trade has been about 60 per cent. A comparison of these two companies makes apparent the differences in sales areas. At Sumitomo Corporation, metals and machines account for 60 per cent, fuel more than 25 per cent, and construction about 10 per cent; the dependence on foreign trade is about 50 per cent. Despite such low dependence on foreign trade, Sumitomo's profitability is said to equal Mitsubishi's.[71]

As is clear from the above, each company has its own comparatively advantageous business sectors. This reflects the particular technological stock and potential of the companies affiliated with each of these trading firms. This is especially evident with Sumitomo Corporation, the number five trading firm, which entered the trading business as late as 1952. The reason it has established a footing so rapidly is that it has been strong in metals, machines, and construction, and it has a large network of production and processing of various materials created through the development of a complex of technology linkages based on the metal- and coal-mining industries, in which various technologies have been integrated and are active.

Since the mid-1960s, the trading firms have been actively recruiting new employees from among graduates mainly from university science and engineering departments. In one company, this type of employee totals more than 20 per cent of the 2,000-strong labour force. This tendency seems certain to continue.

Origins and Functions of General Trading Companies

The general trading companies engage in foreign trade (export and import) and deal with every kind of good, from instant Chinese noodles to nuclear reactors, in staggering numbers of transactions. Today, nine major general trading companies operate in Japan, handling about 50 per cent of all exports and imports (47.4 per cent of exports, 58.6 per cent of imports).

The nine are Mitsubishi Corp.; Mitsui and Co.; C. Itoh; Marubeni; Sumitomo Corp.; Nissho-Iwai; Toyo Menka; Kanematsu-Gosho; and Nichimen Jitsugyo. Of these, Marubeni, C. Itoh, Nichimen Jitsugyo, Toyo Menka, and Gosho constituted the five big cotton-importing companies based in the Kansai area; they specialized in importing raw cotton and exporting cotton yarn and clothes before World War II. The heavy and chemical industries developed after the war, and as they had no experience in foreign trade, the trading companies have become the conduits for the export of heavy and chemical industrial products and the core around which groups of manufacturers have been organized. Nissho-Iwai was established by the merger of

Nissho (formerly Nihon Shoji, which was affiliated with Suzuki Shoten) and Iwai Shoten (a trading firm specializing in iron and steel).

In a look back at the initial stage of industrialization, we find three lineages of trading firms:

1. Those that were incorporated into *zaibatsu*: Mitsui and Co., Mitsubishi Corp., Furukawa Shoji, Kuhara Shoji, and Daito Bussan (Nomura-affiliated). Furukawa and Kuhara went bankrupt during the recession after World War I, and Daito was dissolved in 1920.

2. Those around which a *zaibatsu* was formed: Okura (Okura-gumi), Asano (Asano Shoten), Suzuki (Suzuki Shoten), Morimura (Morimura-gumi), and Iwai (Iwai Shoten).

3. *Zaibatsu* that had no general trading firm: Sumitomo and Yasuda.

Unlike Mitsui and Mitsubishi of group (1), which had their own financial institutions based on their metal and coal-mine operations, the companies in group (2) invested their profits in diverse undertakings, which resulted in the formation of *zaibatsu*, and in the process, the trading firms themselves were transformed into general trading firms (Morikawa 1976).

Nakagawa makes the following observations based on an internationally comparative examination of Japan's general trading companies.[72] Keenly aware of the close interdependence among various industrial sectors, entrepreneurs in Meiji Japan had to make decisions from the perspective of the national economy, beyond the interests of private enterprise, and thus undertook widely organized entrepreneurial activities. The most representative were the general trading companies.

Although trading firms in Europe and the United States were specialized in transactions of specific goods in specific areas, the conditions in Japan were quite different. To compete with these strong enterprises controlling world trade, Japanese trading companies had to start as strong, large-scale enterprises from the beginning.

Secondly, foreign exchange, marine insurance and transportation were well developed in the advanced countries, but not in Japan. The underdevelopment of these auxiliary businesses for foreign trade was an impediment, and so Japanese trading companies had to undertake these responsibilities. To support such activities, the trading companies required a large volume of trade; however, at its initial stage of industrialization, Japan lacked commodities for foreign trade in such volume. The companies were thus compelled to deal in a great number of goods in small quantities to increase volume.

Thirdly, there was a close relationship between the formation of general trading companies and the formation of *zaibatsu*. Each *zaibatsu* made its trading company sell the products *zaibatsu* companies manufactured, which resulted in a transformation of trading houses into general trading companies. On the other hand, some general trading companies moved into the manufacturing industries or established subcontractors there to secure products for transaction, as a result of which the general trading firms grew into *zaibatsu*.[73]

Nakagawa's analysis has relevance for the problems of foreign trade that Japan, as a late comer, had to experience and that the newly industrialized countries now face.

The Japanese Society of Management History made the development of general trading companies the theme for its 1972 convention, which covered not only Mitsui and Co., traditionally the leading subject, but also other trading firms, such as Mitsubishi Corp., Suzuki Shoten, and Iwai Shoten. At this convention, an expert on foreign trade theory pointed out that, even though the old *zaibatsu* have revived as business groups since the war, each grouping of enterprises is now centred round a bank; the Fuyo (Fuji Bank) group, Sanwa group, and Daiichi Kangin group are examples.[74] Where the trading firm, bank, and manufacturers belonging to the same group co-operate fully, the trading companies perform the functions of planning and overall co-ordination.

Moreover, in the process of establishing heavy industry and with growing internationalization, the general trading company diversified the range of commodities it handled. When development projects appeared, the trading company was able to integrate a variety of related goods and industrial sectors, acting as an organizer of industries. Related to this, the trading company has taken on the functions of developer of resources and of collector and analyser of information.

He pointed out further that, during the period of rapid economic growth, when the banks had abundant funds and were careful in their selection of companies for financing under the thriving demands for funds, the trading company, with its credit potential, created intercompany credits and played the role of buffer between the banks and manufacturers. (In this sense, its function is similar to that of the merchant banks in Great Britain.)[75]

Finally, with the ability to raise funds, the trading company increased credit between companies and expanded investment and financing to create a network of mainstay companies and make them its subsidiaries.

While these functions existed before the war, with the increase in the number of general trading companies since World War II, the functions of organizer and information gatherer have come to be more strongly and consciously applied by the trading companies.

At this meeting, Yamamura Kozo, of the University of Washington, Seattle, USA, approached the issue of general trading companies from the viewpoint of economic theory, and reasoned that they (1) decrease the risks caused by the fluctuation of demand in domestic and overseas transactions and by foreign exchange fluctuations; (2) make use of economies of scale; (3) raise the productivity of human and material resources and save the cost of distribution; and (4) enable the effective use of capital and the contribution to social savings.

Yamamura concluded that the realization of economies of scale in the multifunctional general trading companies had saved a huge amount of money and provided the trading companies with the ability to undertake big risks.[76]

Economic analyses of this kind can be persuasive. However, they do not treat the evils resulting from the growth of a trading company into a gigantic general trading company. Mitsubishi Corp., whose male staff used to be referred to as "Mitsubishi gentlemen," had the following incident take place. In 1980, Hokusho, a trading company that specialized in fish products, went bankrupt after buying herring roe at a very high price. Mitsubishi, which did business with this company, was severely criticized by the press for its alleged coercive role in the transaction. Though Mitsubishi explained the circumstances, the president admitted at a news conference in 1980 that, because Hokusho had laid too great an importance on its responsibility to supply the processors, "the company digressed from the principle of market price and did not pay sufficient attention to the end users."

In the period of confusion caused by the oil crisis of 1973, the general trading companies were criticized for their land dealings, stock speculation, and cornering of foods that caused prices to sky-rocket. In 1976, the dirty business of the trading companies involved in the Lockheed scandal was revealed. These evils were evils of general trading firms that have become large and oligopolistic.

In 1974, the Fair Trade Commission, in charge of administering the Anti-Monopoly Act, published a survey report on general trading companies and proposed to establish measures to restrict unfair trading and holding of stocks in other companies. In the following year, a second report was published in which it was proposed to limit bank financing in large amounts. While the Japanese Trade Association, representing the general trading companies, attempted to justify existing trade practices, this new monitoring by the government represented an important turning point for the general trading companies.

During this time too, many analyses and studies were made of the general trading companies. It was determined that, unlike the banks and securities companies of the same tertiary industry, "there is no specific legislation controlling the general trading companies, and there is no such inspection of the general trading companies as that for the banks and securities companies for the protection of depositors and investors."[77] The problem is one of protection of end users.

In the 1970s, the issue of the practices of the general trading firms came to include political and social dimensions. Before that, however, in the 1960s, many predicted a decline of the trading companies. An article by Misonoi Hitoshi was representative of this.[78] By the 1960s, he explains, direct exports by manufacturers had developed, and in the domestic market there had arisen a demand to modernize distribution and thereby eliminate the intermediary profiteering. The development of transport and communications has lowered relatively the merit of the trading company's information-resource function. In addition, diversification of operation and further expansion by the trading companies has increased their expenses and lowered profits.

In the mid-1960s, bankruptcy and forced reorganization were widespread among general trading firms; and yet, despite these problems, there were voices insisting on the need for the trading companies in the light of their changed roles and function in the area of investment and financing in new industrial sectors. At the same time, there was strong criticism from the public in the mid-1970s.[79]

In the 1980s, there has been the opinion in several quarters that the general trading companies are playing a vital role in third world development.

Is the General Trading Company Peculiar to Japan?

Many scholars have argued that general trading companies do not exist in other countries. Such an opinion is evident in Nakagawa's analysis of the general trading company described in the preceding.

Yonekawa Shin'ichi (1983), contrarily, contends that trading firms are a necessary condition in any country for sending as many products as possible to as many markets as possible. He argues that the historical conditions under which general trading companies have not developed in Europe and the United States were peculiar. He claims that he has discovered what could be called general trading companies in Great Britain and in the United States. Yonekawa's argument has opened up a new dimension of the question of the general trading company.

The cases he discovered in Great Britain were the Gibbs Company and the Gasley Company. Both were engaged in foreign trade as their major business; their other activities included marine transportation, insurance, mining, manufacturing, and farming, and they had offices all over the world. The American case was the Anderson Clay Company, which developed from a trading house for raw cotton into a diversified industrial concern.

Arguments of this kind provide good material for further dialogue and studies of any factors that may have inhibited the development of general trading companies in Asian, Latin American, and African countries.

It has been emphasized that the early separation between ownership and management and the recruitment of competent personnel from among university graduates have enabled general trading companies to expand, a notion subscribed to by many scholars. This approach to management differed from that of family-run businesses in the West, where families engaged in related businesses often co-operated for mutual gain.

In the course of the work of this project, it has become apparent that trading firms in other countries could develop into general trading companies under the conditions detailed by our collaborators.

In concrete terms, here are the basic conditions for transformation into a general trading company:

- A business administration that fits the socio-economic and historical conditions of a nation, and a rational, flexible organizational structure (Maeda 1990);
- Keen, well-developed managerial, organizational abilities to cover a wide

range of business activities, from the import of raw materials to processing, quality check, and marketing of finished products. (Suzuki 1990);
- An establishment of a world-wide sphere of activity through a network of branches in various countries, the development of local enterprises or joint ventures to engage in export from or import to the home country, in multi-foreign trade, in technology transfer, in the development of natural resources, and in local production (Kawabe 1990);
- An accumulation of experience through economic development and international relations and the formation of a broad perspective and adaptability (Sakamoto 1990).

Even if the general trading company was established first outside Japan, its soft technology (management) was not transplanted from overseas but formed in Japan under international pressure.

To illustrate, first, there was the trade at the foreign concessions. This was the trade version of the unequal treaties with the Western powers. With financial control in the hands of foreign banks and a monopoly on sea routes by foreign steamship companies, foreign merchants monopolized commercial rights, and Japanese merchants were thus excluded from foreign trade. In addition, there was an arrangement such that, after concluding a sales contract, based on samples, sold items were delivered to the buyer, where they were inspected and measured. A rather high customs charge, called the inspection and measurement fee, had to be paid (this was in force until 1910). When the market situation worsened, contracts were broken by the foreign companies with no offer of compensation.

At the beginning of the Meiji period, the price of exported cotton yarn was beaten down to half or a third its price on the international market, and various taxes, transportation expenses, and handling charges were imposed, all of which meant a big profit for the foreign merchant and a heavy burden on the Japanese. For these reasons, the "recovery of commercial rights" and "direct exports" were the catchwords for Japanese traders, producers, and merchants at the places of production.

In 1881, there was an incident at a warehouse for raw silk in Yokohama. Big traders such as Mitsui and Mitsubishi and bankers such as Shibusawa Eiichi established a warehouse for raw silk. The idea was to secure raw silk for export from scattered small producers, providing them with financing for production, after which the foreign traders were to purchase the silk from these brokers with cash, who, in turn, would complete payments to the producers. The foreign traders rejected this system, and the two sides were in a deadlock for three months. Finally, a lack of funds caused by the accumulating inventory made it impossible for the brokers to finance the producers, and the warehouse was forced to close.

Yokohama Specie Bank, established in 1881, provided credit in exchange for export goods, such as tea and raw silk, that served as security. The government lent a huge amount of money to the bank from the national treasury to encourage direct exports. But the exports stagnated, and some large trading firms went bankrupt. In those days, direct exports could not grow because the trading firms did not have sufficient funds, the banking system had not developed, and the export-supporting businesses were owned by foreigners.

From around 1887, the development of factory production in the spinning industry brought on the need to import a great volume of raw cotton, for which a raw-cotton merchant in Osaka established an importing company. This company, called Naigai Men Kaisha, was consigned by the spinners association to import cotton from China, and it began to bring in Shanghai cotton in 1890. This was the beginning of direct imports. More than 30 years had passed since the opening of the ports in 1858. Two years later, the company contracted with the Tata Company for the importation of Bombay cotton. Another company started to import Indian cotton and Egyptian cotton, and also American cotton, in 1896.

These trading firms soon began to export cotton yarn and clothes to China and became general trading textile firms. This was the birth of the big five cotton traders in the Kansai area. Thus, Japanese trading companies developed with the importers of raw cotton in the vanguard. Likewise, industrial capital was accumulated first in the spinning industry.

In 1876, at the government's request, Mitsui established a trading house for direct foreign trade. But this company imported only woolen cloth for the army and exported only rice, which was delivered to London by a Mitsubishi vessel (the first instance of direct export by use of a Japanese vessel). As it was difficult for Mitsui to purchase and sell in a short period, transactions were conducted with the credit of a trading house that R. Z. Arwin opened in London under his own name; Arwin was an adviser for a company established by Inoue Kaoru, a concern that later was taken over by Mitsui.

In the same year, in concert with government policy, Mitsui acquired a monopoly in marketing coal produced at the government-operated Miike mine (which was later sold to Mitsui) and exported it to Shanghai, Tientsin, Hong Kong, and Singapore. The dual structure of Japanese foreign trade, that is, imports from Europe and America and exports to Asia, started also around 1876 and continued for 90 years, with an interruption during World War II. As I have pointed out elsewhere in this book, coal, especially coal for coke making, and raw cotton became important import items to satisfy the demands generated by the development of industrialization. At the same time, Japan supplied raw materials to nations with processing industries that depended on such raw materials. The trading firms, as the intermediaries in or organizers of both the buying and selling, have become diversified and mammoth.

16

**Industry and Economic Policies—Politics and the Economy
in a New Nation**

The Effect of Economic Policies

The question of why Japan succeeded in modernization with industrialization
as the axis was one of common concern to the participants in our dialogue.
One of our responses to this question was that, because Japan had not been
colonized, it was possible for it to adopt a policy of development and to form
a consensus regarding modernization. However, some expressed dissatisfaction with this answer, as several countries in Latin America had attained
independence early in the nineteenth century. Brazil, for example, was one
of the 10 richest countries in the world, but it was slow to industrialize. The
problem is one of the "speed of industrialization." Thus, it is important to
consider not only why but how modernization by means of industrialization
was achieved. The starting point and process of modernization and industrialization and conditions that determined the structure, form, and pace of
industrialization in Japan were subjects for discussion. The scale and composition of the national economy (resources, the market, etc.) and national
society (population structure and its rural-urban concentration) are also important aspects to study.

Modernization or industrialization, or both, require a series of criteria, not
the least of which is a sovereign nation; others include a high level of social
integration, an efficient administration, social mobility, a market mechanism,
political participation and mobilization, education and information systems,
and a degree of secularization, to name some of the main ones. However,
their mere existence is not sufficient either for modernization or for industrialization. At the end of the Tokugawa period, Japan was in a state of
excessive, saturated equilibrium, which, with a proper impetus from the outside, was able to be changed quickly; without outside force this equilibrium
would have remained.[80]

The effect of continual international pressure on the domestic situation was strong enough to arouse a sense of crisis but not enough to make people desperate: this provides the best condition for the promotion of modernization in a late-comer country. As Sato (1977) has observed, the international environment for modern Japan was generally close to the optimum. And, Japan's geopolitically advantageous position was used to good effect.

The atmosphere of national crisis was heightened by three chief factors, namely, the remote distance from the Western powers and the concomitant uncertainty as to the full extent of the West's military and economic might; China's example in the Opium War—for more than 1,000 years China had been a model for the formation and development of national culture—an example of what could happen when a country rejected the demands of the Western powers to open its doors; and a greater danger of Japan's sovereignty being threatened even by China. Consequently, a national consensus for Japan to model itself after strong nations to cope with them was formed relatively easily.

There has been the expectation in some quarters that, through the generalization and theorization of Japanese historical and particular national experiences and the conditions that enabled it to attain an "industrial revolution," it would be possible for other late-comer countries to initiate their own industrial revolutions. Conditions are different for each country, some created intentionally, others occurring heteronomously; yet, even if we could generalize or theorize, we would not be free from an uneasiness that the conditions we might isolate would not be concrete or effective.

However, that Japan could industrialize proves that other countries can too. For this reason, we established our project to study the development and technology related to each specific industry, to provide positive and negative evidence, a general, inclusive prescription, embracing elements common to all industrializing countries, in order to help these countries work out feasible measures for national development.

An industry-centred approach such as ours has its merits, but it also limits the range and relevance of the problem and tends to fractionate it. For this reason, it was necessary to command a bird's-eye view of all the relevant areas. This was why we set up a subproject to study economic policies.

But why study the age of Matsukata fiscal policy of a century ago? To answer by way of conclusion, the period of the Matsukata fiscal policy was one in which a new nation-state faced a grave crisis. Internally, it was a great turning point. The sovereignty of a nation can depend not only on how military and political problems are handled but also on how the national economy and finances are treated.

A good example of how serious an effect a loss of sovereignty in the economy and finance can have is provided by the fate of the Muhammad Ali dynasty in late-nineteenth-century Egypt. This great power, at the time so much richer and more powerful than Japan, was often said to be the sole power in the orient comparable to the great powers in Europe. Yet it was colonized because of its failed financial policies (excessive loans). Its finances were administered by Great Britain and France, which had adopted conflicting financial policies, the two powers then withdrew their support and Egypt lost all the achievements of modernization it had attained.

Two Opposing Lines, Radical and Conservative

The political change of 1881 (the fourteenth year of Meiji) was the outcome of a confrontation in the selection of policy lines regarding how to overcome the financial crisis faced by the new government. The choice was between a positive measure based on loans (foreign bonds) and a retrenchment policy; the latter was adopted.

From the standpoint of industrialization, this action meant the change from a positive and extensive introduction of Western technologies to the protection and maintenance of traditional technologies (and their modernization). It was also a shift of course from a government-led industrialization policy to the promotion of private businesses, from centripetal to centrifugal tendencies, and a change in emphasis from industry to agriculture. It was also a shift of political power from political-activist-type politicians to administratively oriented politicians. It was a period in which political and ideological confusion yielded to economic and social adjustment. Domestically and internationally, the destiny of the new government depended on such adjustments.

Obviously, the successful treatment of economic and financial problems requires experience and knowledge, and at the end of the Tokugawa era and during the Meiji Restoration in particular, an informed view of the international economy was essential. Fortunately, Japan had a quarter-century after the opening of the country until the consolidation of the Meiji Restoration during which economic and finance specialists could be cultivated. Shibusawa Eiichi, Matsukata Masayoshi, and Maeda Masana are representative. Okuma Shigenobu played an important role in the new government with his expertise in international law. As minister of finance, he handled such internationally important matters as loans from foreign countries initiated by the shogunate government and feudal clans.

Although such talented men were available to the new government, not all were always provided with places and opportunities to play active roles. Though some economic problems were important issues that had critical bearing on Japan's international politics and diplomacy and constituted, in terms of domestic politics, the basis for the stability of the government, they were eventually only part of the whole system of policies.

Each feudal clan had a bureaucracy in charge of economic and financial policies that could exercise a certain amount of power in relation to development problems. Indeed, the policy of promoting industry, the keystone in the Meiji government's economic policy, was an extension, on a nation-wide scale, of what Yokoi Shonan (1809–1869) and others had done successfully in the feudal Fukui domain.

With the opening of the country, both the shogunate government and the various feudal clans immediately lost their financial equilibrium. As studies in economic history have revealed, from the beginning of the nineteenth century, economic and financial problems began to be experienced in each feudal domain. Faced with the dilemma of having to select either an economic development policy that might undermine the feudalistic basis of the social system or a policy of stability through financial retrenchment, the clans individually conducted international economic relations with the various foreign powers.

Each clan had developed its own specialties and a nation-wide market had been formed with Osaka as its centre, and the economic bureaucrats (who were also samurai warriors) had been trained in this system. The opening of the country, however, forced the shogunate to undertake the construction of new port facilities and arsenals and the purchase of warships. Further, the shogunate could not maintain sovereignty with the 80,000 warriors and economic capacity of 8 million *koku* of rice under its direct control, and it encouraged the clans to purchase big ships and other military equipment. This destroyed the monopoly of foreign trade, technology, and international information the shogunate had maintained with its military predominance and its closed-door policy. This also meant an unprecedentedly heavy financial burden for the clans.

The powerful clans in south-western Japan, namely, Satsuma, Choshu, Tosa, and Hizen, overcame the financial crisis, and many of the real power holders in the Meiji government were men from these clans. The Satsuma clan, well known for its benevolent leaders, had the good fortune to have men knowledgeable in fiscal matters during the critical period of the late Tokugawa era, and it was able to transfer the modern technology necessary to reinforce its armaments.

The clan was in the extreme south-western area of Japan, far from the nation's political centre, and smuggling from China and other areas— facilitated by the remoteness of the Satsuma borders—conducted in the guise of trade with Okinawa, also aided the Satsuma clan in achieving its aims. It was through these illicit trade channels that the clan secretly sent students to Great Britain. The clan also had a special young men's organization established to train a political élite to strengthen the clan, and the ties formed among these young men reappeared in later days as personal connections and factions in the Meiji government. (Robert Baden-Powell took the Satsuma clan's juvenile organization to Great Britain, where it later became the Boy Scouts.)

The Satsuma and other powerful clans made full use of the symbolic authority of the Imperial Court to press the Tokugawa shogunate, the holder of political power, to expel foreigners from Japan. In union with power-minded court nobles, the clans gradually drove the Tokugawa shogunate into a corner, and the Satsuma and Choshu clans instigated an attack on foreigners. As soon as they realized, however, the little chance their swords had against the warships of the Western powers, the need of reorganization and modernization of their defence system became abundantly clear.

The reparations for the Satsuma and Choshu campaigns to expel foreigners had to be paid by the Tokugawa shogunate, but the cost was so high that the shogunate embarked on punitive measures against the Choshu clan for its rebellion against the shogunate government. The shogunate also restored government authority in the Emperor and attempted to manage international and domestic political problems through a united regime combining the Imperial Court and the shogunate. There was strong opposition by some of the clans, however, owing also to the shogunate's punitive actions. As a result, those opposed and those not were polarized into pro-shogunate and anti-shogunate factions, which resulted in a bloody civil war that raged for one-and-a-half years.

The battle started in Kyoto, extended to Edo, and finally a portion of the shogunate's forces retreated to Hakodate, in Hokkaido, where the final fight was waged. Through the manœuvring both by Great Britain, which changed its policy from one of neutrality and non-intervention into the domestic problems of another country to one supporting the new government, and by France, which supported the old regime, the fighting in Edo was minimized. That is, to protect their trade interests, both encouraged an end to the fighting.

The Tokugawa regime thus came to an end and a new government was installed. The new government had, as one of its first challenges, the problem of the liabilities of the old one. The total amount of these liabilities exceeded the government revenue for two years, and, except for the foreign liabilities, 80 per cent was written off. The biggest losses were suffered by the wealthy merchants of Osaka. As the notes the clans had issued became valueless, ordinary citizens were also plunged into confusion.

Another difficult problem was the samurai stipends. To transform the old regime into a nation-state, the new government forced the feudal lords to hand over their administrative powers and in their place appointed non-hereditary governors. This shift of power brought with it the burden of continuing the stipends to the samurai class. Although the new government greatly reduced the stipends to the pro-shogunate clans, the total of all payable by the government amounted to approximately half its annual revenue.

In an attempt to lessen the burden, the government recommended that the samurai class give up its hereditary stipend, and offered any samurai who would engage in business payment equal to an annual stipend for six years, on the further condition that he waive forever his and his family's stipend rights. An amount equal to a stipend of four years, partially in cash and the rest by government bond, was offered to those who waived their rights for the duration of their lifetime.

The income for the ex-samurai class thus was reduced to less than 30 per cent of what it had received at the end of the Tokugawa period. The ensuing poverty of this class—which had no means of earning a living—was worsened by the inflation that followed. It was under these economic conditions that the ex-samurai rebellions occurred in different areas of Japan in the latter half of 1870.

The commutation of feudal stipends was carried out by raising foreign loans that amounted to £2.4 million, which was indeed risky. During the age of gunboat diplomacy, Japan had no customs autonomy. Even the income at the Yokohama customs house had been turned over as security for a loan from Great Britain, which was used to liquidate a loan from France.

The new government required feudal lords to surrender their lands to Imperial control and to abolish their domains, which were replaced with prefectures. But many obstacles stood in the way. The lord of the Satsuma clan, for example, had been supporting a campaign by his retainers to overthrow the Tokugawa shogunate, intending to take over the shogunate from the Tokugawas. Despite his contributions to overthrowing the Tokugawa (in order that he himself could assume the shogunate), however, he received no special consideration from the new government, and, not surprisingly, offered no co-operation in the government's reform efforts. Indeed, most of the powerful clan leaders were excluded by the new government, which created a great deal of frustration and resentment, but they were powerless to change the situation.

Furthermore, although there was great dissatisfaction with the government's new policies, the clans' ability to resist had been destroyed by the worsened financial conditions. The leaders in the new government were lower-ranking samurai, men of practical experience who were quite capable of carrying out reforms. In this they differed greatly from the higher samurai class, which was made up of men generally much less competent.

A mere rearrangement of old customs, however, did not consolidate the financial foundation of the new government. For this reason, a land-tax reform policy was implemented. The source of revenue for the shogunate and the feudal clans had been a tax in kind, principally rice. The new government continued this policy, but, because establishing any stable financial programmes based on rice was difficult and because collecting a tax in kind was costly, a policy of collecting taxes in money was adopted.

The government initiated a four per cent tax on privately owned land, to be paid to the central government. After four years of efforts, the tax was implemented on a national scale in 1876. In the following year, the Southwestern Rebellion broke out, fruit of the antagonism that had been building within the political order. This rebellion was the last general rebellion by the ex-samurai class. The war created financial confusion, leading to currency instability and increased inflation.

The government had to cope with the chaos caused by the paper notes that had been issued in addition to the gold, silver, and other metal coins and the inflation that resulted. An effective policy thus required a firm stabilization and unification of the various currencies in circulation. Finance Minister Okuma proposed to solve the problem at one fell swoop by introducing foreign loans of £50 million. It is not known if there was a bank offering such a loan. It might have been a political tactic on Okuma's part, or there might have been a bank with which he was negotiating. In any event, it was, indeed, a bold policy. Once the land-tax reform had been accomplished, the government could undertake financial planning. Though the reform was advantageous, the cash income it brought the government was not effective against inflation (Umemura 1983).

Contrary to Okuma's predecessor, Inoue Kaoru, who was forced to resign following his unsuccessful attempt to persuade his military and education ministers to accept scaled-down budgets, Okuma resigned as minister of finance—a post he had occupied for seven years—because his policy was seen as being dangerously bold.

Matsukata, who succeeded Okuma, adopted a deflationary policy in which a revision of the public accounting system's mechanism of revenue increase and a tax increase were combined. He succeeded in ending speculation in rice by altering the tax collection schedule such that tax payments were due immediately after the harvest. In this way, the financial block was overcome and the inactivity of exports and the attempt to increase imports improved. Thus, the foundation of modern finance was consolidated (Muroyama 1983).

The members of our research team analysed the financial problems faced by the new government from an international point of view, in terms of the global political and economic conditions of the time. This is a departure in the sense that there has been a tendency to exaggerate the value of Matsukata's financial programme; viewing a policy like Matsukata's from an international perspective makes possible a better understanding of what role financial problems (especially foreign loans) play in development and of the importance of maintaining a proper balance between international conditions and internal needs and conditions.

Japanese success was ensured especially by the policy of encouraging industry that was in effect in the period after the Matsukata deflationary policy, which induced a balanced development of the new and old sectors by promoting the development of traditional and local industries.

Maeda Masana conducted a nation-wide industrial survey (in the mid-1880s) and made its results the basis of a development plan published as *Kōgyō iken* (Industrialization policy).[81] He criticized the policy centred on the machine industry as one of the "wrong order," and recommended the development of "native industry." Maeda and his policy line lost in the bid for leadership, and he left government service. He toured various regions, advocating that each village, town, and prefecture should establish its own guide-lines for improvements of agriculture and local industry.

Beginning in the 1890s, the Ministry of Commerce and Agriculture became active in the improvement of animal and plant breeds, of land, and of agricultural technology. The full-scale development of agricultural administration had begun.

As a long-term trend, the Japanese economy began to follow a path of slow growth after the turn of the century. The agricultural industry had been the basis for growth before the opening of the country; due to an extraordinary increase in government expenses after the opening of the country, however, the government was forced to remint coins and accept loans from wealthy merchants, all of which led to serious inflation.

As Japan adopted a silver standard in which its price was set higher than on the international market, importing silver and exchanging it with gold yielded a value that represented sometimes a threefold increase. This brought on a great outflow of gold. The government reminted the coins to match the international standard, but was forced to stop under pressure from foreign countries. In response, the authorities maintained the price of silver, while raising the price of gold by three times. Thus the government effected an international equilibrium at the sacrifice of domestic economic balance. The result was confusion that eventually overcame the Meiji government.

In the period 1872–1884, of the approximately ¥54 million that was reminted, around ¥43 million flowed out and only some ¥11 million remained in Japan. The new government failed in its currency policy, but it was an inevitable reality brought about by the unequal treaties (Nakamura 1983).

The decision on either an overall disposal by means of a foreign loan (Okuma) or a deflationary policy became a political issue. Umemura (1983) states that Okuma's foreign-loan policy had as part of it a cleverly hidden increase of state revenue in real terms of approximately 50 per cent as a result of a currency reform involving a shift to tax payments in cash rather than in kind. Thus the new view arises that deflation had already been progressing in the Okuma era and was accelerated by Matsukata.

In Okuma's policy, inflation was attributed to the sharp increase of the price of gold, and the measure for overcoming it was an overall substitution of currency with specie and the promotion of production on an enlarged scale. Yamamoto (1983) concludes from this that Okuma was an advocate of a policy of promoting industry, but he casts doubt on the possibility that such a large loan could ever have been repaid.

Under rapidly worsening inflation after the Southwestern Rebellion, some influential politicians and businessmen insisted that commodity prices be stabilized by restoring the payment of taxes in rice and by providing the government with the function of adjusting the price of rice. Inoki (1983) points out that the government was reluctant to change to a policy of taxation in kind, as it was believed this would intensify the civil disturbances occurring at the time.

At the time, the land tax provided 76 per cent of tax revenue. To coincide with rice cropping, the tax was scheduled for payment in late autumn or spring. Yet, because expenditures were being made continually, it was inevitable for the system to borrow temporarily more than Y10 million every year. In 1880, these temporary loans amounted to a third of the national budget. Matsukata regarded this as a good reason to lower the value of currency, and he put into effect a minor technical change regarding the period for tax collection that helped bring about an increase in tax revenue and a reduction in expenditures (Muroyama 1983).

A comparison of financial surpluses between the Okuma and the Matsukata periods will verify that they were about the same. In Matsukata's, it was disposed of by transferring the burden to local finance. Although the rice speculation boom, a kind of business cycle, overlapped with the Matsukata deflation, the conclusion is that Matsukata should not have adopted a deflationary policy. Teranishi (1983) states that military and political exigencies forced Matsukata to make adjustments in his policy intention to reduce the scale of the budget.

Local Development

Saito Osamu (1983) has shown how an increasing importance was being attached to traditional industry in the government's policy of encouraging industry, in accordance with its shift from an emphasis on the prevention of imports to one of the attainment of international balance of payments. He also points out that in this process, the local governments had great latitude, although they depended on financing from the central government. In road construction, for example, the ratio of expenditures between the central and local governments was one to two. The Matsukata financial policy provided an opportunity to equalize the shares of the central and local governments in the expenses budgeted for government projects.

Abe (1983) breaks down the changes in traditional industry (e.g. cotton weaving) into (1) revival, (2) decline, and (3) growth, and concludes that a regional division of labour and local specializations were decisive factors. These factors made possible continuous and expanded operations. Enforced uniformity in production was also critical. For example, in Sennan (Osaka), the use of imported yarn for the warp threads beginning in 1875 and the invention of a weaving machine increased productivity two and a half times, and the products of this area outrivalled those from other areas. However, at the time the Matsukata deflationary policy was being implemented, to adjust to a shrinking market, some manufacturers began mixing inferior threads with the imported threads to reduce their prices. This practice eventually damaged the reputation of Sennan cloth, which, in turn, gave rise to the formation of local wholesalers' associations that enforced standards throughout the entire area. Sennan's reputation was thus restored.

An example of a declining trend is Niikawa (Toyama Prefecture). This was a cotton-processing area, and the spinning machine spelled its defeat. The producers in this area were indifferent to technological change and tended to ignore the need to make adjustments to accommodate changing market conditions. In areas that experienced growth, on the other hand, machines were introduced early and the conversion from producing simply white cotton cloth to producing patterned cloth using pre-dyed yarns, in line with a changed market, was carried out successfully. Growth was often characteristic in areas that began business after the Meiji Restoration. Even under the Matsukata deflationary policy, these areas developed new markets in the Kanto and Tohoku areas and continued their growth.

How were these domestic situations linked with the international economy of the time? Japan was already a part of the world business cycle at the time of the Matsukata deflationary policy, and the price of commodities in Japan was affected by world fluctuations. While the world trend was deflationary, Japan was experiencing inflation, which allowed for high economic growth. The period of recovery was accompanied by a drop in foreign exchange linked with the fall of the price of silver. Aided by the growth of exports and by the recovery of overseas business and low exchange rates, Japanese enterprises experienced significant growth.

With the Matsukata financial policy as a turning point, the relations between agriculture and industry changed from one with a leading role taken by agriculture to one of parallel growth.

17

History of Technology and Technology Policy

Politics and Modern Science and Technology

Although Japan has achieved technological self-reliance, it is not playing the lead role in all areas of technological development, and there should be no reason to expect that it should. It is sufficient for Japan to have appropriate areas in which it may contribute to the rest of the world, and any attempt to monopolize the potentials of development in technology would be dangerous. Especially in view of the military potential of the most advanced sciences and technologies, it is important that each country have some share in global science and technology activities, with no one country dominating.

To this end, establishing a world-wide science and technology information network would make an important contribution.

The institutionalization of science and technology has been occurring since World War I, and both in developed and in developing countries, government expenditures to promote science and technology have grown tremendously. In Japan, however, government involvement and investment have been extremely small, the leading role in research and development having been played by private enterprises. In other words, individual enterprises own the most advanced technologies. This is one reason why Japanese R. & D. has often been expressed as "r. & D."

There is the opinion that Japan won the competition for quality control but not for technology innovation. This reflects the difference between countries that have been engaged in developing science and technology mainly for military purposes—for which only high efficiency and quality have been sought and the possible general national economic benefits of them disregarded—and Japan, which has specialized in applying the theoretical achievements of modern science and technology for improving people's lives—in other words, for commercial purposes. It is totally unreasonable to have people dying from starvation in an age when man can go to the moon.

Everyone should be assured of the right to oppose the development of science and technology that has been inseparably connected with such political insanity.

The utilization of the most advanced technology for people's day-to-day living should be developed more positively in each country. Bolder experiments (with certain conditions) should be attempted in developing countries.

However, because industrialization and the development of science and technology involve the principle of private economy, they tend to be accompanied by such negative consequences as those discussed in the section on environmental pollution. Consideration for human dignity and human rights by scientists, engineers, and industrialists must be constantly promoted. As one scientist has described it, it is easy to understand that the most advanced areas related to military technology are the areas containing the most attractive challenges for scientists and engineers. It is most unpleasant, however, to see them exchanging, as an equal trade, the existence of mankind for personal gain.

In this regard, the charter promulgated by the Science Council of Japan in 1980 is extremely promising. The council declared three principles for atomic energy research: the research should be autonomous, democratic, and freely accessible to all.

The charter declares that "science must aim only at enriching the lives of mankind," and "in order to ensure the sound development of science and to promote its useful application," scientists must:

1. be conscious of the meaning and aim of their research and contribute to the welfare of mankind and world peace;
2. defend the freedom of learning and respect originality in research;
3. promote the harmonious development of all scientific fields and encourage the diffusion of the spirit and knowledge of science;
4. be alert to the abuse of science and make efforts to prevent it;
5. respect the international validity of science and promote scientific exchange around the world.

This welcome declaration was an achievement encouraged by recommendations of the Occupation forces intended to fundamentally reform the Japanese system of research and education after 1947 (Yuasa 1984). Just as the Japanese could not prosecute the war criminals on their own initiative, the internal circumstances of late-comer Japan were such that, unless there was pressure from the outside, it was difficult for people to gain the opportunity for change. What is to be noted is that inherent elements had matured so that Japan could respond to pressure from the outside. Under the Meiji state, the potential for reform in the worlds of science and technology had been limited.

Since the Tokugawa shogunate, two historical veins had intersected in Japan, distinguished according to how the idea of "Japanese spirit and Western technology" was used. In one, the idea was used to represent progress, while in the other, it was used to disparage Western technology as merely a tool borne of a culturally and spiritually impoverished world, and, at the same time, to promote a reactionary, exclusive nationalism.

Even as long ago as the ninth century, when Japan was actively importing Chinese culture and technology, there were those in Japan advocating "Chinese technology, Japanese spirit" (Yuasa 1984).

And again, 350 years after the introduction of the gun (1543) and of Christianity (1549) into Japan, the current slogan was "Japanese spirit and Western technology" (Yuasa 1984).

The defeat of China in the Opium War was a serious shock not only to scholars of Western learning but to everyone, leading to the publication of such arguments for coastal defence as *Kaikoku shidan*, a work by Hayashi Shihei. The defeat of China also triggered a shift in the Japanese scholarly community away from the Chinese model, as represented by the *T'ien-kung k'ai-wu*, a seventeenth-century encyclopaedia of technological science by Sung Ying-hsing, to Western science and technology.[82]

In the early nineteenth century, "Japanese spirit and Western technology" had a progressive meaning that continued until early Meiji. However, with the establishment and stabilization of Meiji state power, the nature of the idea began to change, and an abusive manipulation of technology began. The tendency was most marked among military officers and conservative politicians, and there was little public resistance.

According to Yuasa (1984), it was some 50 years before anyone appeared to object publicly and persuasively to the contempt of science and technology that "Japanese spirit and Western technology" was being made to represent.

In 1915, Tanakadate Aikitsu (a science professor at the University of Tokyo) gave an address to the House of Peers (which was led by ultra-nationalists and Shintoists) under the little, "The State of Aircraft Research and Development," in which he campaigned for the establishment of an institute of aviation:

In my understanding, you regard Western civilization as a materialistic, mechanical civilization, a physical civilization devoid of spiritual aspects. You maintain that the Oriental civilization, on the contrary, is metaphysical and spiritual, and that humanity, justice, loyalty, and filial piety are the specialties of the Orient, and your concern is how to harmonize these two currents. However, my question is whether or not we can define Eastern and Western civilizations in such a simple way. . . . I doubt if Western civilization has been established on such a shallow foundation.

Adducing the heliocentric theory of Galileo, he continues:

Once he had concluded that the earth revolves around the sun, he resolutely maintained his conviction. . . . Such an attitude cannot be taken to be from intentions to make money, attain fame, or the like. Perhaps this is the mind that lodges behind what is called the materialistic Western civilization. It seems to me that there are many living Galileos and Newtons in contemporary Europe and America, and they are cultivating the source of civilization.[83]

In this, we can recognize, beyond the digressions peculiar to Japan, the establishment of modern scientific thought with a universal nature. Nonetheless, this kind of thinking was not victorious over the transformed notion of

"Japanese spirit and Western technology." Those who supported the military fascist regime, which was interested in Western science and technology only as a means, were aware, on the one hand, that the outcome of any modern war was a matter of scientific and technological potential, but, on the other, as a serious contradiction, rejected the "ideas" inherent in modern science and technology, thus finally leading to self-collapse.

Unfortunately, there was a revival of the pre-war notion of "Japanese spirit and Western technology": the new version held that Japan's loss in World War II was a defeat of its scientific and technological capabilities. Post-war reconstruction of Japan was therefore to be aimed at Japan becoming a great power through development of its science and technology.

The conflict here is an antagonism between the nature of politics and the philosophy of science and technology. The relationship between science and politics has reached a critical point in the present age. Tanakadate cited Newton and Galileo, who lived in a time when science was not yet institutionalized, was politically independent. The moral difficulty of scientists in more recent times was well represented by Einstein. As a result of the institutionalization of science, however, modern scientists now constitute a part of the power élite, and one would be hard put to find an Einstein among the agrochemists who developed Agent Orange or the economists who discuss the kill ratio of a particular weapon and their like.

These people are unaware that they are "fierce animals" bred as specialists of knowledge and technology within the state system of the United States. They are "specialists without spirit and sensualists without heart." The contemporary age is one of a gigantic institutionalized science. An analogy may be drawn from the mafia chieftain who is also a devout church-goer, activities that in his consciousness are non-contradictory. This unconscious schizophrenia has debilitated modern scientists and engineers.

There have been criticisms and warnings from the scientific community against the induction of top scientists into the bureaucratic power structure, and therefore I need not take the matter up here. Among Japanese scholars, Hiroshige Toru and Nakayama Shigeru have published excellent works.[84]

In this regard, we can be sympathetic to the warnings and hostilities of the participants in our dialogues concerning specialists in science and technology, although their criticisms are stereotyped and lack concreteness. The privileges and influences of the élite in science and technology are as great in developing countries, in which the absolute number of scientists and engineers is small, as in the superpowers. In such circumstances, there is no other body of scientists in these countries that could act as a counterbalance, a check to those who become a part of the power élite. What is worse, those élites continue in office even after changes in government.

Technology Policy for Development

Nakaoka (1986), in tracing the history of modern technology in Japan, points out the surprising similarities to development in the West. Indeed, while

Japanese technological development has been unique, it constitutes a part of a world-wide trend in technology development, and therefore no country need follow rigidly the particular developmental pattern of any one country; rather, each should choose the one most suited to its special needs and conditions.

Focusing on the Japanese iron-manufacturing and cotton-spinning technologies, Nakaoka concentrated on management, labour, supporting technologies, and related services. He examined Japan's lag in comparison with the development stage of iron-manufacturing technology in the West when Japan started and how it caught up with the West. He emphasizes that the existing large stock of skills and accumulated technology in Japanese traditional iron-manufacturing was of critical importance, in spite of the fact that the scale of production was small and equipment was primitive.

At Kamaishi Ironworks, for example, a shortage of charcoal forced the plant to switch to coke for fuel. The attempt to manufacture coke failed, however, due to an incomplete understanding of the manufacturing process, and it was not until management was privatized that success came: the scale of operations was reduced to balance the supply of fuel with demand, and the links with transportation and other services were rationalized.

Nakaoka emphasizes the critical role played by the establishment and adjustment of linkages with such peripheral factors as fuel, power source, and transportation problems in determining success or failure. He also observes that the time required for each stage of development in the West was considerably shortened in Japan (a total of 400 years versus about 50), "as if the Buddle blast-furnace had actually been skipped over."

Nakaoka also examined the history of the modern development—after passing through the three stages of its early development—of the textile industry, which confirmed the importance of related technology and supporting services. He also traced two significant transformations: (1) from management by administrators to management by business professionals and the formation of skilled labour and (2) from the employment of farm labour, which is completely unaccustomed to the systematized, regulated working conditions of a factory, to the employment of factory workers.

Nakaoka stresses the significance of fringe technologies in spinning as the crucial elements leading to Japan's industrialization success. The technology for manufacturing wooden machines, for example, was developed by the makers of looms and water-wheels, who had high-level engineering skills. Native technology was joined with imported Western technology to form an intermediary, or hybrid, technology. The mechanical systems, key parts, and energy sources were Western, and they were combined with native Japanese parts etc., to create the necessary linkages and stages in the production process. Toyoda Sakichi, for example, invented the automatic loom using imported cast-iron parts for the key components, while the rest was constructed of wood. Toyoda was able to duplicate the metal English loom by cleverly combining local materials and know-how with those from the West.

Ishii (1986) makes reference to a kind of technological linkage in which a new technology does not replace traditional technology; rather, what is fostered is a coexistence of mutual prosperity through the realization of a supplementary relation whereby, taking cloth-making as an example, the spinning is done by labourers at the factory, while the weaving is done by part-time workers at home. In this example, the most difficult and labour-consuming processes were performed by modern machines in the factory, while the weaving process could be done at home in the traditional manner because the Japanese market demand was for narrow cloth, suitable for use in the making of the traditional kimono. Western weaving machines were capable only of producing broadcloth, and so they were installed in the factories for weaving cloth to be used for military uniforms and cloth for export to other Asian countries.

Incidentally, the development of cotton spinning placed cotton ahead of linen among textile materials. In the present age of chemical fibres, on the other hand, goods made of cotton or linen have come to be considered luxury items.

Regarding technology policy, Uchida (1986) examines the history of technological policies from 1825 to 1935, dividing it into four periods. He described the characteristics of each as follows.

The first is the period up to the Meiji Restoration (1825–1868), when the prototype of the Meiji government's technology policy was to be found in the policy of the Tokugawa shogunate and other feudal clans. This was also the time when absolute government control over technology and information was crumbling as a result of both internal and international events. The promotion of industry by feudal clans as a financial policy to increase cash revenue would form the basis for the new government's central policy of industrial promotion on a national scale. Since the establishment of the Tokugawa shogunate, the samurai class was evolving from one made up of warriors to one of administrative bureaucrats, which aided the government in its formation of an administrative apparatus and the development of needed economic and technological policies.

The second period (1868–1885) was characterized by impetuous Westernization represented by model factories under direct government operation in which not only were the equipment and machines imported but the engineers and foremen were also brought over from abroad to provide operational guidance. In 1873, the number of such foreigners totaled 239 (from 1875, the total number began to decrease). In this period, however, the government had no comprehensive technological programme or technology policy. Opinions were divided among leading politicians, and each government office was engaged in its own programme of technology importation and personnel training, with no integration at the level of the central government.

In the third period (1885–1910), there was an important policy change: for the first time bureaucratic management of state-run factories gave way to private management, as a direct result of accumulated debts and fiscal difficulties of the central government. The change of policy was accompanied by a movement of personnel from government enterprises to the private sector, which provided an opportunity for a broad dissemination of technology. On the other hand, in such areas as railways, meteorology, and communications, which were kept under government control, each ministry established its own school to train and secure human resources. Laboratories, research institutes, and experiment stations were established for agriculture, industry, and fishery technology in this period.

The technology policies of the army and the navy were characterized by a strong inclination for weapons independence (home production, uniformity, and standardization). This goal was attained by the beginning of the 1910s (though later for the navy, because of the nature of its weapons).

Self-reliance in technology was achieved mid way through the fourth period (1910–1935). At this time, the minimal linkages among technologies had been established on a national scale and a new stage of development then began. Though in the past military technology and science had been strictly in the hands of the government, in this period, further technological development required the participation of the private sector. The military's policies aimed at Japan becoming a superpower corresponded with the government's goal of making the country a first-class industrial nation.

In this (last) period of tremendous development of the heavy and chemical industries, the machine and chemical industries, under the momentum provided by World War I, were able to substitute their own products for imports, thus bringing to a halt their costly purchases of many goods from abroad. Also in this period, newly-risen *Konzern* formed big business groups with the technologies of the heavy industries as their central pillars. One such group was the Institute of Physical and Chemical Research, which was transformed and enlarged from the Institute of Basic Science Research, established with government assistance in 1916.

Uchida concludes that it was only with the establishment of a technology agency in 1942 that technology became an independent item in Japan's total national policy.

Addendum

Concerning post-war science and technology policy, the establishment in 1959 of the Science and Technology Council was significant. Its members include the prime minister (chairman); minister of finance; minister of education; director-general of the Economic Planning Agency; director-general of the Science and Technology Agency; chairman of the Science Council of Japan; and three others appointed by the prime minister. This represented an unprecedentedly powerful system for the administration of science and technology. Prior to the establishment of this council, the Science and Technology Agency (established in May 1956) had been active in carrying out the administration of science and technology, aiming at the promotion of science and technology to contribute to the development of the national economy.

According to Yuasa (1984), the two 10-year programmes established by the Science and Technology Council, that is, (1) the "Basic Measures for the Promotion of Science and Technology Aiming at Ten Years Hence," of 1960, and (2) the 1977 "Science and Technology Policy in the Age of Limited Resources—A Basic Ten-Year Programme," were extremely significant. Though monotonous and long, the government documents describing the programmes provide us with a good picture of the problems Japan faced.

The response by the Science Council of Japan to the 1977 proposal was summarized as "Science and Technology at the Turning Point" (1978). This and the "Scientists' Charter," published by the Science Council of Japan in 1980, make a pair. It is noteworthy that in the charter, Unesco's "Recommendation on the Status of Scientists" was interpreted as both a "charter of rights" and an "ethical code."[85]

In a study, "The Historical Development of Science and Technology," conducted by the Policy Research Group and chaired by Sassa Manabu, the call was made for a new, holonic and flexible path of scientific and technological development in view of the observation that there was a trend toward overspecialization (atomism) in science and, at the same time, the creation of Big Science.[86]

18

Female Labour and Technology Change

Total Household Labour and the "Wife's Domain"

Household labour is the main form of labour in societies based on agricultural production, but the household in Japan was not necessarily a unit based on blood ties. Depending on the scale and type of production, outsiders were incorporated into the household to complete a single production unit, and all members, regardless of age or sex, were expected to work together. Thus, even the heaviest labour was performed by both men and women. What sexual divisions there were resulted from the women undertaking tasks that men were unwilling to perform. Female labour was indispensably important, not only in the preparation of meals but in the production, care, and storage of foodstuffs and clothing and in social activities centred on festivals, weddings, funerals, and similar events. Child rearing was not counted as work.

Thus, the situation in the traditional, non-farming Japanese household, in which production and social relations are the male domain and housework and child rearing the female, was not the norm in farm households. Yanagita Kunio (1875–1962), the father of Japanese ethnology, stated that the peasantry had to have more than 100 different skills to be able to provide the variety of high-quality necessities that formed the basis of a self-sufficient life-style. And these skills had to operate within the limits set by nature, beyond any human control. Besides the basic handling of crops, tasks needing to be done included the making of fuel, fertilizer, sacks, and rope; house repairs; and jobs essential to the basic functioning of the community, such as road repairs and maintenance and supervision of temples and shrines. The peasant was thus not only an expert agriculturist, he (or she) was also at times blacksmith, carpenter, stonemason, earth mover, hydraulic engineer, veterinarian, hunter, woodcutter, and weaver. Ideally, the peasant household had at all times someone able to function in one or more of these capacities.

Because the size of the household has shrunk over the past century, one peasant household after another has fallen into ruin, unable to keep enough members to maintain their existence. During the same time, new and smaller households have appeared, family units whose existence has depended on the multiple efforts of the wife.

An increase in the number of family members had been thought desirable in principle to enhance the labour force, but the bitter experience of many families was that too many children were "more a burden for the present than a help for the future."[87]

A new problem, requiring a new solution and a new setting for its solution, arose as the limits of agricultural production were reached amidst an unhealthy environment of high birth rates and high death rates. The new, smaller families could not hope to find a solution to their predicament either in the village or in agriculture. From the beginning of Meiji, surplus labour left the villages, and the basic production unit—all family members working together—was transplanted to the towns. A large family in the countryside was thought to be proof of surplus wealth, just as a small family meant poverty. In both extremes, all household members had to work extremely hard, and whether families remained in the countryside or not, all were defined by the need for all to work together to keep the household viable. This was the Japanese household when the Japanese economy was beginning to take off.

Yokoyama Gennosuke, author of *Nihon no kasō shakai*, a report on labour conditions at the end of the last century, points out that, in a survey of 1,615 Tokyo establishments employing 50 or more workers, 111,913 were men and 184,839 women (for a total of 296,752).[88] Even if we take for granted that women workers would outnumber men in such light-industry factories as those producing raw silk, tea, and matches, not to mention the weaving and spinning factories—this was after all an age of light industry— Yokoyama also looked ahead, to an age when "we will have to depend on the machine for production and women will increase in even greater numbers among the work-force."[89] He pointed to the decrease in the numbers of male workers in lantern, spinning, and camphor factories.

Further, "even in such industries as mining and iron and steel production, where one would not expect to see women working, one in fact encounters quite a few," Yokoyama noted, recording that more than 300 females worked daily at the Tokyo Arsenal.[90] He explained, however, that most of these women were family members of other arsenal workers.

The need to make ends meet compelled entire families into the factories. "In Japan, only glass, shoemaking, and metal-working factories relied on skilled labour"; the day would come when "we rely entirely on machines," and, consequently, the number of women workers will increase.[91]

P. H. Douglas (1902–   ), using wage data, proved that the size of the household budget determined the amount of outside labour a household would supply. In Japan, Arisawa Hiromi provided a similar analysis.[92]

According to Douglas and Arisawa: (1) the lower the income of the head of the household, the greater the chance that family members will work outside the household; (2) when the income of the head of the household is stable, the chances of other family members working outside increase with better pay; (3) the household head will seek work without regard to the level of wages. This theory, based on the premise of female labour supporting the household income, is persuasive and widely applicable.

If we combine this theory with Yokoyama's observations, we find that the appearance of female and child labour in the market-place was due to changes not only in technology but also in the household economy. Examining Japan's experience, one is better able to understand why European skilled workers made such a negative response to technological change and why labour unions and left-wing political parties opposed the participation of women.[93] The response to the crisis of the traditional view of the family and the household economy was negative. In Japan, however, the response to the capitalist transformation of technology was made by entire households based on the co-operative, whole-family labour practice characteristic of the peasant economy, perhaps befitting the nation's status as a late industrializer. And this situation has not changed much today. Of families in the 1980s depending on wage labour for their livelihood, "more than half have both husband and wife working, in one form or another. . . with wives of low-income earners especially liable to be working and contributing a sizable portion of the household income."[94]

Times have changed since Yokoyama made his observations. The standard of living and consumption have risen to an entirely new level, but wages have risen only after prices, often after a great time lag. Following the oil shock of the early 1970s, stagnation has affected real disposable income. This represents a modified application of the second part of the Douglas-Arisawa theory: female labour has increased, principally as part-time labour, which implies an increase in low-paid work.

Whereas before the oil shock renovation in technology brought women into the market-place to fill a labour shortage—and, in the case of married women, to defend their own household budgets against inflation—after the oil shock, companies selected part-time female labour as a cost-saving alternative. This has been made possible by the technological streamlining effected in factories (FA, or factory automation) and in offices (OA, office automation).

To understand the place of female labour in Japanese society, it is necessary to look at the wife's position within the household economy. The male head of the household in Japan does have ultimate responsibility for fiscal management (if there are finances to manage), but, traditionally, and this holds true today, the budget and the actual running of the household are the responsibility of the wife, who receives all monies brought home by her husband. After consultation with his wife, the head of the household receives an allowance from that income. It is rare for the husband to interfere in her management of the family budget, which, incidentally, partly accounts for the fact that the nation's leading economists are ignorant of everyday prices in the market-place.

It is true that a trend among young working couples is for both partners to manage the budget, but these people are still in a distinct minority. The wife's control over the family budget has no relation to the size of the income, but where all must work to maintain the economic viability of the household, her authority increases dramatically. Her responsibilities then expand to include not just all the housework, even though she is no longer a pure housewife, but also to the earning of a portion of the household income through labour in the market-place.

The biggest driving force behind supplementary family income is the cost of housing, especially of home ownership; the "new poverty" exists in housing. Because of the high cost and limited availability of housing in Tokyo, most commuters must spend an average of 90 minutes travelling each way between home and job, and even then they cannot afford a house with a garden.

The next greatest demand on the family budget forcing wives to work is the cost of children's education. Through the past century of political, social, and economic upheavals—continuing in the present drastic technological revolution—what has proved to be of most lasting value is the knowledge, qualifications, and skills acquired through education. Thus there is a high regard and high expectations for education. For the middle class, into which the majority of the Japanese fall, an educated citizenry with income but no wealth, to survive in an urbanized, industrialized, highly technological information society, families can pass on but one thing to their children and grandchildren: education.

Still following the Douglas-Arisawa theory, an increase in the income of the head of the household results in the belated entrance of his children into the labour market, after a period devoted to education. The serious burden the housing problem imposes on the family budget makes it unavoidable that the wife also enters the labour market.

I thus speak of the survival, even the thriving, of the "wife's domain," even though there is no legal basis for this authority. For the Japanese, this acknowledgement of the special role and domain of the wife is an ethic that functions subconsciously within everyday life, a natural part of that life. For this reason, especially in a large household with two or three generations sharing a rural domicile, it is usually thought natural for a bride to spend long periods of unpaid labour learning work and management skills, even under what may seem to be unbearable work conditions, to support the family. In this capacity, the bride is functioning as a candidate to be the mate of the head of the household, her husband's acceptance depending in part on her performance.

Technological Change and Female Labour

Throughout the world, as long as new technologies failed to generate the development of newer technologies or of new branches of industry, the effects were limited to the pre-existing branches of industry and few new jobs were created. However, the spread of new technologies raised productivity, developed new products, and, by means of lower prices, deepened and broadened the market, so that employment at least was not reduced. In cases where prices were not lowered or the market expanded, however, the spread of new technologies did generate reduced employment.

One characteristic of modern technology is its tendency toward skill saving. But, when an economy is in a stage of development and expansion, technological change does not exclude skilled labour. Instead, the saved skills are spread to new, previously unskilled labour. In general, even if modern technology causes a reduction in the number of skilled workers, specialized, high-level skills are still necessary; indeed, their importance is enhanced. Knowledge gained from experience, based on previously existing high levels of education and knowledge, varied skills, and expert judgement become critical assets. The skill-saving process involves a parallel reduction in time needed to acquire new skills and an increase in less-demanding work. This process coincides with the rise of female labour as a percentage of the total number of workers employed.

A simple way to demonstrate this process is to consider the changes in the proportion of labour employed directly in production with that employed in ancillary fields. In Japan in 1950, the ratio of men employed in the former was 1.26 times what it was in the latter; for women, the ratio was 3.1. Most women were employed in agriculture. By 1970, however, the ratios had changed to 1.53 for men and 1.06 for women; the 1980 figures were 1.15 and 0.61. Total employment for each year was 32,020,000 in 1950 (13,940,000 females), 52,110,000 in 1970 (20,390,000 females), and 55,650,000 in 1980 (21,070,000 females).[95]

The most distinctive change in these figures was the decline of female employment in production and its movement into ancillary fields. Where three women had once been employed in production for every one working elsewhere, by 1980 these three were matched by five in other fields (a ratio of 1 to 1.7).

As a result of technological change during this period, the skill saving of labour and the rise of the service sector in the economy brought on changes in the structure of female employment. Among men, in contrast, although skill-saving and the service sector's rise have affected the production branches of industry, most male workers remain engaged in productive labour. Women have moved from agricultural work to office work and the service sector (represented by an 80 per cent drop in agriculture and an increase of 3.6 times in the clerical and service sectors).

If we look at wage differentials between men and women, female wages amounted to 42.8 per cent of male wages in 1960, 47.8 per cent in 1965, 50.9 per cent in 1970, 55.8 per cent in 1975, and 53.8 per cent in 1980. The shortage of labour in the period of rapid economic growth caused wages to rise, and the continuing presence of females in production work also caused the gap to shrink between their wages and male wages during that period, but after the oil shock of the early 1970s, the economy entered a period of stable growth, and the gap again began to widen, if only slightly.

On this macro-economic scale, distinct differences are evident between male and female wages, but on a micro-economic level, where jobs require special skills, qualifications, length of service and training, the difference disappears. We must therefore look for wage differentials between men and women in non-skilled, non-specialized jobs. Also noteworthy is the phenomenon of frequent changes in place and type of work among women. Female labour is not abundant in the kind of permanent jobs that hold a central place in the labour market. Female labour is fluid and peripheral.

This last point is related to the proportion of women actually emloyed in the labour market relative to the total population of women at least 15 years old. In 1980, 47.6 per cent of all women were employed. This represented a slight increase over the post-war low, reached in 1975, when only 45.7 per cent were employed. The recession caused by the oil shock brought on a reduction in female employment. The proportion of women engaged in productive employment in 1975 dropped below the proportion of those in ancillary occupations. Also in 1975 the national unemployment figures topped 1 million for the first time in the post-war period, and, of this figure, 340,000 were women, who, with older men, were a buffer against depression.

The highest proportion of employed women is in the 20–24 year-old age group; the lowest is in the 25–35 year-old age group. After age 40, the proportion again rises (55 per cent of urban women in their 40s are employed; 70 per cent of rural women in the same age group). The M curve for Japan is the sharpest among all industrialized nations. It coincides with the withdrawal of female labour from the labour market during the period of early marriage, childbirth, and child rearing. Because the peak of employment is in the 20–24 year-old age group, the proportion of 15–20 year-old women in school is high.

Compared with other industrially advanced countries, Japan's lack of nursery and day-care facilities stands out dramatically (especially the scarcity of centres able to care for children the entire time parents are commuting to good-paying, full-time jobs). On the other hand, relative to the underdeveloped countries, the burden of housework is much less severe, and the nuclear family has become dominant.

If the extended family is characteristic of the Asian household, we can say that in the post-war period the Japanese household has rapidly become non-Asian—that is, nuclear—as a result of rapid industrialization and urbanization. In 1980, the total number of households was 34,080,000, with the average household comprised of 3.3 persons. Even in a country such as India, the extended family system is said to be maintained only in wealthier households. In 1946, in Japan, the average household size was 5.09 persons, though that figure reflects the special conditions and devastation resulting from the war. In 1920, the average was 4.85, which remained more or less unchanged for some 50 years.

The proportion of female labour in the entire work-force was 38.2 per cent in 1920. Considering that it was 37.9 per cent in 1980, the labour market has experienced long-term stability. The number of household heads has not risen, nor has the percentage of women workers. What has changed, though, is the pattern of work, with a shift of female labour from the primary to the tertiary sectors of the economy. Over the long term, this reflects technological progress and improved female education.

Focusing on the primary sector of the economy, no significant lightening of the work burden or spread of services is detectable. Instead, the mechanization of agriculture has increased the burden of female labour. To pay for new farm equipment, the head of the household must seek employment off the farm. This has become common practice among purely farming households, and most men must travel far from their farm homes for work. The work they find is the kind of menial drudgery or night work the urban dweller shuns. As one young farmer summed it up, "We must increase productivity to ease our work load, and we need money for that; to get that money, we must leave the farm. Something is wrong."

This is often referred to as the poverty of modernization. Since the machines needed for increasing output and productivity are so expensive, work must be sought far afield to pay for them. Consequently, all of the non-mechanical work in agriculture is becoming the province of female labour. To allow the farm wife to devote more time to farming, labour-saving household machinery has spread to the farm.

When scholars from underdeveloped countries visit the Japanese countryside, the usual reaction is an insistence that, with their radios, televisions, cars, Tokyo newspapers delivered at the house, Japanese farmers are not farmers—they want to see real farmers. What they fail to notice is the occurrence, in the women especially, who shoulder a huge burden in having primary responsibility for the farm and family, of the same kind of ailments that afflict factory labour.

Family labour was once the basis of Japanese agriculture, of its efficiency. Now the Japanese countryside is characterized by the same kind of "both-partners-working" situation that has long defined the lower-class urban couple.

In fishing villages, a different effect of mechanization has emerged. The spread of small motorized craft has destroyed the old way of life in which the wife processed and sold the fish brought in by the husband. Now what the fisherman needs, since he can go farther out for his catch, is a partner to help with the net spreading, hauling, and carrying. The wife's role in this is made possible by the lightening of her housework. Family income has risen, to be sure, but so have expenses and debt, which have necessitated harder work by the fisherman and his wife. The intensification of family labour in fishing mirrors the same process once characteristic of the economy's agricultural sector.

These changes are a result of the urbanization and factory-type proletarianization of labour. This "urbanization of life-style," for example, the spread of home appliances and the products of the factory—food and clothes—has aided in the lightening of the burden of housework.

Whether this structural change in the way of life represents a qualitative improvement is another matter. Nevertheless, the change is probably irreversible and it creates new problems of life-style and culture. Indeed, there are those who lament that, though people were poor before, they at least had "lots of good things to eat." Even given that taste is partly a matter of custom, universal standards remain; consider the widespread popularity of French or Chinese cuisine.

The quality and content of the Japanese way of life are now being transformed by the overwhelming dominance of computers and the electronics industry in the national economy. Previously, the heavy and chemical industries had this effect. Whether we view these developments positively or negatively, the rapid changes taking place in the structure and content of family life cannot be denied.

These changes, whose effects have reached into the primary sector, have both broadened the scope of female labour and enlarged the problem of female unemployment. The new employment opportunities being offered are available only to those women with proper qualifications and sufficient education who are thus able to respond to the changes in technology. Labour that is unable to adjust to the changes is shunted aside. The results are lower wages and worse working conditions, a spread of part-time work (with less cost for the companies), and indirect hiring. There is thus the problem of women abandoning the search for work in the highly technological labour market-place. This problem is especially common among middle-aged and older women, creating an economic crisis for the "all-work-together" household and, indirectly, for the culture that has maintained the institution of the wife as household budget manager.

The particular structure of labour and the unemployment of middle-aged and older women in Japan are a result of the original decision Japan made to industrialize. The decision was not simply a matter of applying Japanese values; it was the result of the particular historical experience of the Japanese people. Modernization is not a problem whose solution can be postponed or drawn out, but one that requires a once-and-for-all solution. It is a question of national society and culture, and the problems each country faces cannot be blamed on others. The efficiency of Japan's farmers in response to the demands of modernization assumes new importance in this consideration. We must also recognize, however, that conditions prevailing when Japan began its modernization, both external and internal, were markedly different from those faced by underdeveloped countries today.

Although little attention has been given them, I contend the two aspects of complete family employment and the wife's domain, her role as the household budget manager, were important contributors to the favourable conditions prevailing when Japan undertook to modernize. These two features characterized the farm and small independent family-run businesses.

This position may be criticized for overstressing what is simply an ethos, a generalization of a particular life-style, the life-style of the independent producer, for whom it is natural for all family members to work and for the wife to assume great responsibilities within the household. The Douglas-Arisawa theory, which describes the later stage of Japanese industrialization, makes apparent that the social and economic characteristics of the independent farm household were transplanted to the urban setting, so that it can be seen that the above argument is not merely an after-the-fact characterization of capitalist society in its early stages. Economic history and the history of technology show that the current problem of female unemployment is a result of the present social and economic stage of industrialization, which is one characterized by the mass consumption and production of diverse products manufactured in small lots.

The term "wife's domain" is based on Yanagita Kunio's coinage *shufuken*. My own use of it is aimed at deepening the possibilities for discourse on the Japanese experience, with reference to the problem of underdevelopment. In Japan, 70 per cent of labour is externally employed (versus 90 per cent in the United States), leaving as much as 30 per cent that is self-employed or family labour. It would be an obvious exaggeration to say that only this 30 per cent of the population nourished and preserved the vitality of Japanese society, but without a doubt, the notions of communal work and the wife's domain are most appropriately applied to this sector of the working population. The ethos of a wife's special responsibilities regarding household finances permeates the other 70 per cent of the population, accounting in large measure for the willingness of housewives to work once their child-rearing responsibilities are reduced.

At this time, however, this pool of labour cannot be fully absorbed or employed, as reflected in the unemployment figures for women, which are significantly higher than for men (in 1980, 3.1 per cent female; 2.6 per cent male). Since the concept of unemployment varies from country to country, international comparisons are difficult. The Japanese concept includes no one until they have entered the labour market, which explains the lower figures than for other countries.

Nevertheless, there is a serious problem of structural unemployment in Japan, one neither easily grasped nor easily solved. High levels of female unemployment occurred both in 1970 and in 1980, with the latter being marked by a spread of unemployment among middle-aged and elderly women as a result of the automation of office work through the introduction of labour-saving devices. Japan thus exhibits a trend—which is opposite the experiences of other industrialized countries—towards a more youthful labour force.[96]

Addendum

In this section I would like briefly to return to the early years of the Japanese spinning industry, highlighting a few of the leading figures and works of the period.

Upon leaving the government-owned Tomioka Spinning Mill, Wada Ei (1857–1929), along with her co-workers, was awarded the special title of

"Women Spinners' Victory Battalion." The year was 1874; such magnanimity in a factory supervisor would not be possible only a few years later. This incident provides an expression of the spirit of early Meiji, the days of "a rich nation and a strong army," of government sponsorship of industrial development.

Later, in 1913, Wada wrote that "military men would doubtless have been furious at this act, but given the national effort that the government coordinated in building this factory, such an expression of feeling seems only natural." The year 1913 was also the year her husband, a professional soldier, died from wounds received in the Russo-Japanese War. Several years later, Wada began her famous *Tomioka Diary*, composed while she was living in the Furukawa Mine company house at Ashio Copper Mine. Although she most likely never met Tanaka Shozo (1841–1913), the famous leader of the struggles against the severe pollution from the Ashio mine, coincidentally, 1913 was also the year in which he died, in the nearby village of Yanaka.

Wada Ei's life—a spinner while still a young girl, officer's wife, and mother of a mining executive (she died during her second stay at Ashio, in 1929)—represents the course of modern Japanese history. Her *Diary* was written in 1927, but it was not published until 1931, after her death, by the Shinano Educational Association, in Nagano Prefecture.

Another work about the spinners, *Jokō aishi* (The sad history of the girl spinners), written by Hosoi Wakizo, was published in 1925, one month after the author's death from acute peritonitis. These two records of factory life, one from early Taisho and the other from late Taisho, represent "extremely valuable material." This was the evaluation at least of Wada's book by the Nagano Prefecture supervisor of factories, Ikeda Nagayoshi. It is likely that Ikeda knew also of Hosoi's work.

In the summer of 1927, when Ikeda made this comment, a strike of unprecedented scale occurred in one of the largest factories in the Nagano spinning area, at Yamaichi Hayashi-gumi, in the city of Okaya. After the third week of the strike, the employers resorted to a lock-out, closing of the dormitory, termination of meals, and the use of hooligans and the police to suppress the strikers. The 1,300 strikers suffered a bitter defeat.

The first signs of recovery from the 1929 depression began to appear in 1931, and this was when Japan began its move towards a war-time economy. Also at this time, the Shinano Educational Association chose *Tomioka Diary* to be a school reader. What was the reaction to the *Diary*, and later to *Jokō aishi*? Since the events of the strike were probably still fresh in the minds of many readers, their reactions must have been varied indeed.

The threat of financial ruin was a powerful force in the early years of the textile industry. The term *seishigyo*, which means "spinning industry," could also, as a pun, be understood to mean something like "life-and-death industry." This word-play aptly symbolizes the life-and-death struggle constantly waged in the textiles market-place. As Nakamura (1985) has pointed out, the good times of 1919 quickly turned sour in 1920, forcing many establishments out of business. Managerial strategy to deal with the crisis took two forms: buying up of cocoons and wage cuts.

According to Yamamoto (1952), what would often happen was that wages would be reduced or recalculated, so that part was held over until the next season. If a spinner did not agree to work the next season, however, this portion "often would never be paid." So to be sure to be able to receive the deferred portion, the spinners would stay at their jobs, which was just what the deferments were intended to induce.

Around the turn of the century, a carpenter's wages were about 60 sen (1 yen was equivalent to 100 sen) per day in the rural areas and 66 in Tokyo; this was at a time when rice cost 119 sen for 10 kilograms.[97] A high-ranking spinner was earning more than 100 yen per year, an ordinary spinner in the range of 40 to 50 yen (according to records from 1899). This was indeed, in the words of the popular phrase, the age of the "100-yen spinner." One acre of paddy around Hida, in northern Gifu Prefecture, cost between 100 and 150 yen at the time.

However, only a determined minority of skilled spinners were able to earn such wages: one in four was said to be in debt at the end of each year. But wages, which were stable between the late 1880s and the mid-1890s, began to rise rapidly towards the turn of the century.

Not long before this, however, there were many cases in which spinners were "not paid anything that could be called a real wage," receiving instead used clothing or cloth as compensation. This was common among Hida girls working in Nagano; in nearby Hirano (present-day Okaya City), ordinary spinners were paid wages comparable to the very low wages of road-gang workers.

The spinning factory owners, motivated by a sort of Protestant work ethic, drove themselves and their employees—many of whom were mere children—mercilessly. Their sober enthusiasm for hard work, which they forced on their workers, was, at the same time, a manifestation of the cruelty of the capitalists. Holding wages to within 5 per cent of total production costs and buying from a wide variety of cocoon suppliers to provoke competition and thus keep supply costs down, were means to cope with the wide fluctuations in the silk market.

Because of the importance of a stable labour supply, and because differences in spinning skills made for great differences in product value, owners made great efforts—including attractive wage rates and a bonus system—to retain their more-skilled workers.

In both the silk-spinning and cotton-spinning industries, there was a need for skilled machine operators, and so, many highly skilled women spinners were able to use the chronic shortage of skilled labour to their advantage, travelling around the countryside and working at different factories and thus creating for themselves a relatively free existence. But others, as described in *Jokō aishi*, underwent great trials, such as those who were permanently blacklisted for union activities. This happened to several friends of Hosoi, including his wife.

"Spinner for ten years, waitress for a year and a half, five years on the black market, twenty years as a day labourer, and twenty as housewife . . . no house, no pension . . . too old to work . . . . " This was how Hosoi's wife rather modestly summed up her life. Though her life contrasted greatly with that of Wada Ei, their lives overlapped, forming a counterpoint in women's social history. The reflections of Hosoi's wife, published in 1980 as *Watashi no jokō aishi* (My sad spinner's tale), read with Hosoi's book and alongside the *Tomioka Diary*, provide a unique glimpse of the spinners' lives and hardships. Another work of interest in this regard is Hayashi Fumiko's autobiographical work, *Hōrōki* (Diary of a vagabond).[98]

These and other works are not well known outside Japan, and it is hoped the brief mention of them here will stimulate interest for future dialogue.

19

Industrial Technology and Pollution

"Development" and the Destruction of the Environment

When spring comes and there is no sound of birds singing, we are made to recognize the occurrence of drastic ecological changes that threaten the existence of all life forms, including the human race.

In her 1962 book *Silent Spring*, Rachel Carson provided an early warning of the danger posed to the delicate balance of life by the unrestricted use of agricultural chemicals. Hers was a splendid, if disturbing, critique of our civilization.

During the same period, the seas around Japan were threatened and the nation's quality of life also endangered. Writer Ishimure Michiko was living in Kumamoto Prefecture's Minamata, site of Japan's worst pollution case. Quoting one of the victims of Minamata disease, she writes, "The fish are a gift from heaven. They are something that we take, free of charge, when we need them, as naturally as life itself goes on. Now where can we turn? I would pray to heaven, but heaven itself is ill."[99]

Carson depicted the deformation of inland areas and Ishimure witnessed destruction along the sea-coast. One could add the words of an elderly resident of a coal-mining town: "We were all poor when I was a child, but we had good things to eat on the mountain. Nothing grows there now."

Here we have one result of industrialization, which, up to this point, we have counted as a positive thing, but, as is obvious, it has negative consequences too.

Even in developing countries—indeed, there especially—industrial pollution is a huge problem. The bitter experiences of the developed countries can serve to warn the developing countries.

Formerly, pollution and environmental destruction were produced by industry, and this is still largely true today. But now the primary sector of national economies has become either the source or the proximate cause of pollution, making the problem serious indeed. This is because "development" is first of all a response to the population explosion, and its first priority must be to increase the food supply. To this end agricultural chemicals (pesticides, herbicides, fertilizers) have been developed and applied to raise food production. This results in what Carson calls the beginning of the destruction of the delicate balance in the chain of life.

There is the added problem today of atmospheric discharges resulting in acid rain, which destroys forests in countries thousands of kilometres away from the source, kills the plant and animal life in their lakes, and pollutes their seas. Acid rain is thus an international problem, and not all of it stems from large factories.

Because of the increase in the sources of pollution, even though each source may meet emission control standards, their combined long-term effect mean an enormous and absolute increase in pollution build-up.

And thus it becomes difficult to identify the polluters. It becomes impossible to assign responsibility for the pollution that results from the total discharge, since each source may in fact be meeting the established discharge limitations. There is no legal basis for assigning responsibility or guilt in these cases. In the mean time, environmental damage and the danger to human health are on the rise as a result of these discharges.

When the polluters also suffer the effects of pollution and when their numbers reach a certain level, the problem becomes unsolvable by legal means. The legal system operates on the presumption that the violators of the laws will be a minority and that the majority of the population will co-operate to maintain the system.

Thus, the environmental problem is becoming a political problem, requiring political initiative and a newly defined (broadened) legal conceptualization. The environmental problem has grown beyond the capacity of the present legal system. Despite the limitations, however, environmental disputes are being pursued along conventional legal lines, and, from this effort, a new concept of the issue in terms of human rights has arisen. However, because it has been difficult to assign a monetary value to the right to a clean environment, the legal system has not been able to deal adequately with the problem, and thus the inefficacy of the legal system becomes apparent and the environmental problem becomes politicized.

As mentioned, industrial pollution and environmental damage are not limited to the industrialized countries. They are also serious in the third world. The desertification of central Africa is due to the population explosion and the increase of livestock raising. The forests of Thailand have fallen below the level of 40 per cent of total national land area deemed necessary to preserve their reproductive capacity. Land erosion and flooding in north Thailand have resulted, spreading damage as far as Bangkok. The air in South American cities is so bad it makes travellers from Toyko ill. The year 1985 will go down in history as the year of the two great human disasters that befell Mexico and India.

Minamata disease, thought to be peculiar to Japan, has reportedly appeared in Iraq and in north-east China. There are also reports of its occurrence in Finland and Canada. There is a time lag between the occurrence of pollution, environmental damage, and society's recognition of the problem. How long that lag is depends on that society's view of human life and human rights.

The problem of acid rain has been discussed in Europe for a long time now. Its solution is difficult because of the great distance between the source of the discharges and the areas they damage, and this is compounded by the international nature of the problem. Yet damage is increasing without regard for particular political or economic systems.

For this reason, the pollution problem must be addressed from the viewpoint of the victim, in terms of the basic human right of survival.

Pollution and environmental damage are the by-products and the ill effects of development. Once there were cries of "Give us pollution!" from members of the élite in some third-world countries, showing the high priority development has had and continues to have in those countries. Since surveys are not carried out and even surveys by foreign experts prohibited by some governments, it is difficult to know just how serious the environmental problem is in these countries.

Arguing against development, however, is not the answer; those who oppose development are in effect telling millions of people to starve to death.

What is important—and feasible—is discussing how development is carried out. Development is a right of every nation, the key to the formation of a modern nation. But no nation has the right to carry out development without regard to the threat it may pose to the basic human rights, basic human survival, of those in other nations. As a citizen of the nations of the earth, I must advise against over-emphasizing development to the detriment of the lives and welfare of every planetary citizen. One citizen may not threaten the life or property of another, and governments should not be permitted to do what citizens may not do.

Certainly pollution and environmental problems make the task of development more difficult, but they must not be ignored because they do. Instead, an international effort must be mounted, using fully the experience and wisdom of all nations.

In this sense, however directly applicable Japan's experience in other fields may be, the "Japanese experience" is most directly applicable to other nations on the environmental level.

Japan is advanced as both a polluter and a producer of antipollution devices. But the latter comes as no honour, since even the most advanced antipollution technology cannot erase the accumulated damages of the past.

The Question of Diagnosis

Even in medical diagnosis, where assessment is much easier than in environmental pollution, serious legal problems exist regarding assigning responsibility, tracing causes and linking effects.

Modern medicine is capable of diagnosing a great variety of diseases, but it tends to consider only particular, individual symptoms, thereby missing the larger picture and failing to grasp the totality of new diseases resulting from pollution. By separating individual symptoms and analysing them as if they were not related, the totality of the problem is completely ignored, and symptoms resulting from combined effects are improperly understood.

In the conventional method of diagnosis, test results showing mercury in the hair and in the blood confirm mercury pollution. Yet even this recognition leads to the rather scattered diagnoses of pulmonary obstruction, asthma, diabetes, high blood pressure, arthritis, and kidney damage, without recognizing that the actual culprit is Minamata disease. This has happened in Finland, Canada, and China.

Minamata disease was defined by researchers at Kumamoto University as being caused by the presence in the body of certain levels of mercury: 200 ppb in the blood, 300 ng/l in the urine, and 20 mg per 50 kg of body weight.

Harada Masazumi has pointed out (1985) that this definition is both vague and inadequate. It deals only with acute typical Minamata disease and does not cover non-acute, late, incomplete, or fetal exposure to the pollutant. By thus defining the disease narrowly, the authorities were able in 1960 to declare that the disease had been eradicated.

The Kumamoto University research group looked at only the initial symptoms of Minamata disease, and of these, only the more acute ones.

Harada, in contrast, insists that an "epidemiological" method of diagnosis be employed in diagnosing Minamata. By this method, victims and non-victims "would be compared as groups, and their group health trends" closely examined.[100]

This methodology is a function of modern statistical analytical techniques. Factors that cannot be examined in an individual analysis of patients, such as environmental factors, can be placed in their proper context and defined over time, through space, and in terms that also include all other relevant factors.

It is this point—that diagnosis that is impossible in individual cases becomes possible on a group level—that is so important in regard to the methodology of diagnosing environmental destruction and its effects on human health.

The point, evident in the example of Minamata disease, is that, just as the problems of the environment and pollution cannot be properly understood in terms of pre-existing theory, neither can the new diseases be diagnosed by means of conventional diagnostic practices.

An understanding of a disease involves an appreciation of a great many related factors, since the ultimate aim is to prevent initial damage and, if that cannot be accomplished, to aid the victims who have been harmed.

Even when the polluter and the source of the pollution can be identified, the polluter may not have the resources to compensate for the damages, further complicating the problem. In such cases, the government, although perhaps not directly involved, is compelled to step in and bear some responsibility for solving the problem. And as a result, it tends to minimize the problem in order to minimize its responsibility. This is what happened after the outbreak of Minamata disease; its definition by researchers at Kumamoto University effectively cut off from assistance everyone whose individual symptoms did not fit the narrow terms laid out.

Here again the problem becomes one of basic human rights. Everyone involved in working toward a solution to the pollution problem is involved because of a concern for basic human dignity. An unavoidable loss of that dignity occurs when specialists concerned with the problem of pollution focus narrowly on "professional interests"—be they those of the technician, the lawyer, the doctor, or the bureaucrat—to the exclusion of an overall solution.

The narrow view of duty to one's profession and unquestioning obedience to one's superiors that marked the handling of the Minamata disease case brings to mind the attitude of the man responsible for the deaths of so many at Auschwitz during World War II, Adolph Eichmann, who also had no inclination to disobey orders that violated the basic dignity of humans.

As for the victims and the protectors in the Minamata and other similar cases, people stood up for human dignity in a way that self-serving bureaucrats and professionals overly concerned with wealth and status would never do. The courage of the victims and their supporters brought renewed hope.

The Prototype of the Present-Day Pollution Problem

Technology is necessary for development, and technology hastens that development. But that same technology, even when applied in the proper way according to existing standards, can lead to unexpected damage and victimization. It can also lead to a very severe worsening of natural disasters. In these cases, however, the responsibility of the polluter is all too often unclear.

Pollution in Japan first appeared in connection with mining. Mines create their own pollution but they also contribute to natural disasters and intensify the effects of environmental destruction for which they are not the direct cause.[101]

Because the pollution problem occurred in a strategic industry, the protests of the farmers whose fields were being polluted and the pleas to close the mine were ignored by the authorities, and when the protests grew, the army was brought in to suppress them.[102]

For its own sake and to meet the needs of the military, the Meiji state pursued development with little regard for human welfare; a concern for human rights and the environment was sorely lacking and all pleas and protests were suppressed. Indeed, in many cases production was accelerated, causing further environmental and health problems.

However, even if protest and resistance can be pushed aside, pollution problems and damage remain. In the end, even heavier costs must be borne. Such has been the case in Japan many times over the years.

Forced development to satisfy the needs of the state has also occurred in

Canada, where members of its native population now suffer from Minamata disease.

As governments compete to lure corporate investment, that competition weakens concern for protection of the environment. Since corporations do not operate on the basis of societal needs, and since they will invest where there are profits to be made, even if environmental controls are strict, they will stay. They will leave, even if controls are lifted, when profits decline. If we rush to promote development at all costs, we risk underestimating the behaviour of technology owners, the corporations.

When national development needs and big business's interests match, environmental problems are extremely difficult to solve, considering the size and strength of state power. This is true also of nationalized industries. It is a myth that there are no pollution problems in the socialist countries.

Summarizing the Japanese experience, from the perspective of pollution, we can say that the post-war period of rapid economic growth was also the peak period of pollution, resulting in protests and campaigns to have the problem addressed. And it was addressed in the development of new anti-pollution technology, in the introduction of various controls: anti-smog controls are stricter in Japan than in any European country. The responsibility of polluters has also been clearly defined. Companies that pollute, it is safe to say, will not survive.

This does not mean, however, that Japan has eliminated all pollution. According to experts, the controls are effective, but there is still a slow, steady build-up of pollutants. Also, as anyone can recognize, the decline in the level of pollution is not unconnected with the slowing down of economic growth, with the depression of the manufacturing industries.

The economic downturn occasionally makes people nostalgic for the days when pollution was pouring into the environment. It is a dangerous illusion.

One elderly person in a ravaged, abandoned mining town had this to say: "We were all poor once. There was not much to eat, but what we had was good. Well water, mountain herbs, river fish: it was all there for the taking, and it was good. Perhaps those were the real luxuries."

Today's developing countries need to industrialize and to raise their food production to feed their growing populations, but we would not have them choose to destroy nature, which provides, free of charge, things so good but so precious they can be called luxuries. We would urge them to re-examine conventional technology, to substitute, where possible, alternative technologies.

In the development of new concepts of technology, it is desirable to minimize the use of chemical controls, and in planning this, the people of the areas affected should participate in decision-making for development. If not, it will not be possible either to use the accumulated wisdom and experience of the local populace or to carry out smoothly any plans developed.

The benefits from development are great. They come as surely as the damages resulting from it. Even a temporary dislocation can prove disastrous for people already on the margin of existence. Any assessment of the advantages and disadvantages of development should take into account the warning of the founder of the Club of Rome, Aurelio Peccei (1908–1984), who urged developers to ensure safe development not only during our lifetime but for the coming generations as well.

Medical Myths

Harada Masazumi, the Minamata researcher, in noting how the discovery of Minamata disease destroyed old medical myths (e.g. the foetus is in no danger so long as the mother is healthy, since it is supported by the nutrition of the placenta), points out that mercury poisoning "threatens the foetus even when it does not harm the mother." Eugene and Eileen Smith brought the plight of 40 foetal victims of Minamata disease in Japan to the attention of the world, but there are many others in Sweden, America, and Iraq.

This discussion of pollution has not covered all aspects of the problem of pollution in Japan; it merely touches on the cases so thoroughly documented by Ui Jun and his colleagues who participated in our project. One final point to be made concerns the presence of occupational diseases in companies that pollute, showing that there are intimate links between such diseases and pollution of the environment.[103]

Part 3

Epilogue

20

Conclusion: A Proposal for Future Research

Before coming to our proposal, I wish to reiterate that what I have introduced above does not represent all the work of my colleagues on this project.
There is that of Otsuka Tsutomu and Kasama Aishi on food-processing technology; Nakagome Shozo on Western clothing manufacturing in Japan and
traditional fabric recycling technology; and Namie Ken on farmers' development activities. Two other regional studies were planned, but were in the
provisional stage when this book went to press; there is every reason to
expect they will bear fruit.

Time and budget restrictions prevented us from publishing the results of
the work of these scholars. Ironically, problems in "developing" these new
areas of study were encountered. Although the work in these areas is of high
practical and scholarly interest—indeed, because it is so—we find having to
leave this work unfinished or unpublished particularly grievous.

We were also perhaps overly desirous of as wide an audience, as large a
number of "dialogue" partners, as possible. This pushed our staff to the
limits. But we did our best. Our respective places of employment all suffered,
all bore the burden of this project. And all deserve a thankful mention here.

In conclusion, although we were not able to achieve all that we had hoped,
we are satisfied that we have contributed in some measure to the narrowing
of the gap between theory and practice. I fear only that I may have let down
those working with me on the project.

The "Japanese experience" project was an attempt to analyse the process by
which modern Japan passed from dependence on foreign technology to technological self-reliance as an example of a national experience with the development problem.

This perspective has heretofore not been found in Japanese studies,
whether in Japan or abroad. As a result of our five-year study, we have
developed our own theory concerning the components and stages of develop-

239

ment. Here, in place of a conclusion, I would like to put forward a specific proposal regarding the problems of technology and technology policy and the course of future study. As stated throughout this book, what we desire is dialogue with third-world intellectuals and practitioners now actively engaged with the problems of development and of technology in the context of development. This proposal could serve as a basis for further dialogue towards a solution of these problems.

In exchanging actual experiences, specific examples, we participants in the dialogue can gain a clearer picture of development and its problems and identify common elements and possible solutions, and the proper priority or scale of the urgency of problems and how they might be linked. Through such a dialogue, I will doubtlessly be compelled to revise my position on certain aspects and go back to the shop-floor to provide my colleagues with the information that may be needed to give them a fuller understanding of the Japanese experience, to help them solve or provide them a way for tackling their particular development problems.

Specificity is desirable in the debates surrounding the problems of development and technology, especially reference in the debates to specific historical experiences. Without this specificity, there can be no approach to a solution. We cannot afford the privilege of leaving the solution of these problems to a third party. The problems require detailed inquiry. But detailed inquiry alone is not enough. Unless we converge specific cases to generalize, there can be no methodological basis for our dialogue. We cannot say, based on convergence from specific examples to theory, how far the dialogue we are urging will develop. We have, after all, only limited experience with such dialogues. Nevertheless, we must expect that our efforts will help to lift the discussion to a higher level for those who come later.

The Role of the State

1. This subject fits in anywhere in the dialogue. Looking at the Japanese experience, we find dramatic differences between the structure and the operation of the Japanese state between the time of initial industrialization and the period after World War II. The locus of sovereignty was also different. For these reasons, we must be cautious about simple or excessive generalization.

It is easier to discuss the role of the state if we confine ourselves to governments. The government played a major role in national development in Japan, which in a sense took place both at the initiative of the state authority and in the cause of its further development. That it was intended as protection of national sovereignty is obvious from the government's policy towards railroads and landholding patterns along rail lines. The government simply invoked its authority to acquire needed land resources.

At the same time, the government invoked its sovereign prerogative to build government factories and offices, for which it imported foreign technology and employed foreign engineers to bring needed technology into the country.

But the government did not do everything. It lacked customs autonomy and was unable to apply a policy of protection for infant industries. This was due in part to the small size of its tax base and thus of its budget. Because of budget difficulties and also the failure of government enterprises to prosper (not to mention the government's over-reliance on foreign engineers), many government-operated factories and mines were sold to private entrepreneurs. As a result, technology began to spread and management began to be rationalized.

The railroad enterprise, begun at government initiative, demonstrated its efficacy and profitability, leading to private rail development on the local and regional levels. The government example proved the efficacy of development.

Others learned from government failures. This was the case in the spinning industry, which was able to find a "rational" scale, in terms of technology and management, for proper operations after observing the government's mistakes. Here again, the government initiative performed a useful role and did not act to contain private activities.

In general, we can assert that, while the government bore the main burden in heavy industry, iron and steel, armaments, the main rail lines, and communications—industries requiring large capital outlays—it left the development of light industries to private entrepreneurs. By pursuing this kind of policy, it was able to foster the transplantation of the machine and chemical industries and to nourish the growth of a national technological network. This network had both public and private aspects, but the state was able simultaneously to minimize its own role and to maximize the efficacy of that role.

This did not lead to autarky in technology, but instead to a rapid and varied transformation of technology. If rational choices are made in the importation of basic technology it will be possible to integrate the imported technology into the pre-existing technological stock without creating too great a burden on the national economy. Japan's army and navy, contrarily, aimed at developing technology for "weapons independence," stressing functionality, with no regard to cost and linkage. Furthermore, it was not possible to apply such technology in the civilian economy, as the technology was developed with purely military purposes in mind.

In conclusion, we can say that the Japanese government lacked a firm policy on science and technology in the early period and began to develop one only after the outbreak of World War II. And, not until 11 years after the end of the war did the Japanese government establish its Ministry of Science and Technology, in 1956.

2. Japan's ability to develop technology, in spite of its slow response to the challenge, was due to the low level of the West's technological development in the mid-nineteenth century and to the technological gap between Japan

and the West being much narrower than what now exists between the North and the South. Japan was lucky.

World War I caused a structural change throughout the world in the relationships between politics and science and technology and also between science and technology and society. It also produced wider gaps between the West and Japan in certain sectors of production. Having achieved the minimum necessary level of scientific and technological autonomy, however, this period represented a challenge that Japan was able to meet. In time, and after persistent analysis, Japanese manufacturers were able to replicate the latest technological developments of the West and thus produce products that were competitive and narrow some of the sectoral gaps between Japan and the West.

Taking the latest technological developments as models, Japan copied the leading technologies, borrowing and replicating creatively. This approach, after the interval of World War II, provided a basis for Japan, once economic recovery began in earnest, to do what the scientific and technological leaders had done after World War I with national policy for science and technology. Japan's second stroke of luck was that this development came when the trading of scientific information and technology was freer than it had ever been.

The guiding principles of this Japanese experience can be represented by the following key words: the *feasibility* needed to realize technological goals in a relatively short period, the need to be *selective*, the *strategic planning* involved, the requirement for a clear establishment of *priorities*, and the desirability of *orderly* and *timely* application.

The effective application of these principles requires governments to secure adequate information about the five Ms discussed earlier and the five stages of development from imports to self-reliance.

Today, both science and technology in their most advanced areas have become heavily military in orientation and are very large in scale. A high degree of national and political prestige are involved, and efficiency and functionality have been realized to a degree far beyond the demands of mere economic rationality. Science and technology have thus become politicized, a part of the establishment, and in this sense, totally unrelated to the problem of development. Even if certain technologies derived thereby can be used for development, they will demand such economies of scale as are quite impossible for developing countries, handicapped by diseconomies of scale, to provide.

Further, contemporary technologies are, in a manner of speaking, black-box technologies in that they are not amenable to the kind of disassembly and study the nineteenth-century technologies were and not easily integrated with indigenous technologies, also often possible in the nineteenth century. For these reasons, if technologies are not borrowed selectively, their importation can easily lead to technological subordination and an end to hopes for autonomy.

In today's world, no country is autarkic in science and technology. And only a few have the ability to develop new technologies. If countries without this capacity choose technologies that worsen their technological subordination, these choices cannot be said to have been advantageous to the development of the nation-state.

3. Granted that development is a function of sovereignty and that technology is necessary for development, countries must nevertheless establish firm technological goals. Once these guide-posts are set up, the role of the state is reduced to monitoring how they are adhered to. Here again, administrative ability becomes important. This is something in which developing countries are often deficient. Data and information necessary for the establishment of guide-lines are not just neglected; they are left to rot in storage or appropriated for private use. These organizational weaknesses are compounded by poor monitoring, so that when something is really needed, it is not available. This last problem will probably be remedied when the latest information technology is employed.

From our dialogue, we have learned that there are countries with no industrial census, and that many of those that do have one restrict its use. A census must be taken, and proper methods for its use established. Unless the level of accuracy in industrial census data is raised, technological planning is impossible and guide-line monitoring ineffective.

Unification and standardization of units of measure, of rail track gauge, and of electric voltage and cycles are necessary, for example, to the developing nation. Countries lacking this unification face major obstacles in their pursuit of development. This may seem a minor point at first, but for countries that were kept divided regionally and culturally when they were colonies, such a logistical unification is often very difficult to achieve. Once these standards have been firmly established for technological development, however, enterprises can be set up fairly easily, and the newly urgent need to modernize outdated equipment and plants will spur national standardization.

As the record of village activity contained in Namie's (1981) work shows, one factor inhibiting the spread of knowledge of technology (and the application of that technology) is the use by techno-scientists and bureaucrats of terminology and units of measure that are unfamiliar or difficult to understand. This was true in Japan until the 1930s, especially concerning fertilizers, but generally evident in most books on agricultural technology. What we found from our surveys and dialogue was that there were surprisingly few basic technical manuals available, and the effort to provide charts, tapes, and slides for the illiterate farmer and to broadcast needed information for the spread of technology was noticeably lacking.

4. A technology and science policy has several aspects. First is the overall approach to development. The dicta "universal truths" and "science as the answer to everything" need to be carefully scrutinized to see just how applicable they are. Also, a policy on science and technology must include a persuasive response to those antiscientific and antitechnological elements in society that will resist its implementation. Taking the long perspective, which

involves a conviction of the efficacy of science and technology, developing nations must first take advantage of being late comers to industrialization. Next, developing nations must establish industries which are appropriate to their particular cultural conditions and seek competitive advantage in order to gain competitiveness in the international market.

Second, a policy of science and technology must be a policy of industrial technology. First designate the locomotive industries of the economy; then, based on their technology, develop linkages with related industries and supporting service industries.

At this phase, the problems directly encountered by a developing country vary widely from country to country. The technology for solving specific problems of development often does not exist anywhere, and each nation must develop its own. Technology transfer can only help to solve these problems; it cannot provide the basic solution. What is often called the latest technology is not necessarily applicable to the problems of all countries.

Reversing the explanation, we can say that the "alternative technology" is whatever technology can solve problems of development, be it new or old, domestic or foreign, acquired easily or developed with difficulty. The important thing is the creation of a national system for the use of alternative technologies. For self-reliance in technology, this must be created as rapidly as possible, at the minimum scale and standards required.

Third, countries wishing to develop a sound science and technology policy should plan for the training of scientists and engineers.

This comprises two basic strategies: (1) a long-term strategy for developing specific standards in elementary, middle, and high school education and (2) a short-term plan for meeting more immediate manpower needs. Both are related to the need for the creation of a cluster of engineers able to work with old and new, domestic and foreign technology to bridge the gaps in linked technological development.

Science may be universal, but technology is not. Native engineers must be the creators of solutions to problems of national development.

National Consensus and Basic Human Rights

Friction cannot be avoided in development. This is because there will inevitably be those who benefit and those who suffer. Even if it were possible to conceive of a situation in which this were not so, there is no historical precedent for development that preserved social relations and the social structure and that also brought about social and economic development that was both horizontal and vertical. Even when a proper balance is struck, there are bound to be differences in time or across regions or among different groups in society. If there are not too many people divided into opposing camps of beneficiaries and victims of development, the many who fall somewhere in between or who are neutral may act as mediators in the conflict.

Because development is intended to benefit a nation as a whole, when individual regions or groups do not share in the benefits, cries for fairer distribution can arise. The resulting tension and social unrest can have harmful consequences for modern technology. Because of the high density of interrelated technologies, paralysis or disorganization in one key sector (for example, in railroads or electric power) can have widespread and lasting effects.

A basic requirement of modern technology is political, economic, and social stability. At the same time, development and the proper use of technology promote stability. Meiji political leadership was remarkably stable and was therefore a basic condition of the domestication and development of modern technology in Japan.

However, it cannot be overemphasized that a national consensus supported the importation of modern technology for development. It was through the creation of this consensus that the social tension and friction resulting from the process of development based on imported technology were minimized. It was precisely because of this consensus that the accumulated experience and wisdom of the nation could be mobilized effectively and applied to the solution of problems of development.

A national consensus helps to absorb the shock of development, but it can also work the other way, as happened at the Ashio Copper Mine. The people living near the mine had their basic human rights and livelihood ignored and violently and forcibly suppressed. Tanaka Shozo, the leader and organizer of the protest movement against the pollution caused by the Ashio mine, was politically destroyed by the Meiji state and died in poverty.

Although not widely known internationally, the significance of Tanaka's work and thoughts is great indeed. He was not a modern thinker; yet, for that very reason, he was in a position to directly criticize the negative aspects of modern technology. Tanaka was not against technology or development *per se*, but to how modernization was carried out and how it affected people. Tanaka maintained that some groups in society were victimized, and that there was a need for the majority to respect the basic human rights of the minority. Tanaka's was a criticism from below of the national consensus for modernization that was initiated by the élite; it was a criticism of the inevitable disrespect for human rights by the authoritarian national statist ideology of the time.

To express it differently, we might say that, although a national consensus supporting development must be created, as long as it ignores the problem of human rights, it will never be free from the evil of state authoritarianism, that modern disease.

Because development is a matter of national prerogative, there can be no denying a country's right to it. Similarly, final responsibility for resolving the confusion and friction arising from development lies with the state. Government activity is legitimized by national consensus and regard for basic human rights. The basic human rights we speak of need no strict legal definition. They involve human dignity, the right of each individual to a decent life. There is a tendency for conflicts to arise between required technology and basic human needs, between human needs and those of the state. In times of political or economic hardship, the needs of the state tend to come first. It is the duty, and, if accomplished, the honour of politicians to co-ordinate a respect for and protection of basic human rights with the national interest.

The government imports and supervises technologies for social development and the establishment of a national economic base. There are also technologies that private individuals and groups use in the development of industry. When the two kinds are linked, they can be structurally co-ordinated, and the basis for an autonomous national technological infrastructure can be created. Once established, this autonomy can later be rapidly extended. The sign of a truly sovereign government is the early establishment of structural links between the technologies the government needs and those private industry demands. Many governments have asserted their sovereignty and tried to protect their economies, but few have thus proved sovereignty.

Formation of a National Technology Network

Modern technology is linked and accumulative. Its vertical and horizontal structural links, compared with those of pre-modern technology, guarantee less freedom of manœuvre, despite the normality of free transfer and sale of technology as a commodity. Without a variety of stable, fixed-scale service industries and a supporting infrastructrue, modern technology simply cannot operate.

This is the reason even technologies transferred for a specific productive purpose (for example, for import substitution) can quickly run aground.

Today, some technologies are transferred to meet government needs, and some are transferred for the needs of the populace at large. Hitherto, most technology developed in Europe and the United States has developed from the bottom up, over time, to meet the needs of civilian industries. Because the development was relatively slow, society had time to deal with the tension and friction arising from it. Although in Japan's case development was from above through government leadership, in contrast to the European pattern, and in this sense represents a case of late development, Japan also followed the classic pattern of first developing light industries as a domestic basis for the shift to a heavy industrial economy. Thus, Japan never faced the truly serious problems confronted today by developing nations; its situation was less complicated. This is not to say, however, that Japan had an easy time of it, that it could develop without bearing heavy burdens and without great suffering.

Because Japan experienced a combination of modernization and industrialization from above with technological accumulation and development from below, the stage of creation of a national technological infrastructure—the minimum necessary to sustain national autonomy—was largely finished more than 60 years ago.

Japan built up its primary technological infrastructure in 60 years. This was possible because it had the five Ms necessary for technological development and an accumulation of traditional technologies that, while only partly useful, were still transferred in part to the modern sector. This forging of organic links between old and new, foreign and domestic technologies made possible the achievement of technological autonomy, the creation of a national technology. Making full use of artisan skills and the accumulated wisdom of premodern society can make possible the copying of complicated machinery, especially when even vaguely similar types of technology have existed earlier. Examples of the role of these traditional technologies can be seen in Japanese clock-making, sword-making, and castle-building technologies. And yet, even if such skills prepared the way for creative copying, they alone were not sufficient to sustain modernization and industrialization.

It was Japan's engineers and creative entrepreneurs who were able to combine proper information and evaluation of latent skills and technologies with modern technology in a smooth course of development.

This makes clear the need—indeed, the duty—of a developing nation's engineers and its founding entrepreneurs to rediscover and to re-evaluate the value of that nation's traditional technology if a truly national technology is to be developed.

The promotion and orientation of research and development constitute an important aspect of this evaluation; unfortunately, in the cases known to us, no developing country has made a proper inventory of its native technology.

These latent links are most important for agricultural technology and its related technologies. The population and food problems have reached unprecedented proportions in many countries. It is easy to see the importance of developing agricultural technology, but the problem is complicated by the need for links between old and new and foreign and domestic technologies. Secondly, social and industrial development must be linked with agricultural technology, especially in the areas of fertilizers, water control, and the building of a national road network.

However, if the people living in the areas that will be affected do not participate in the planning for development, maintenance and supervision will be difficult and deterioration will soon begin. In Asian countries where the hydraulic facilities are more modern than Japan's but where the peasantry has no say in their maintenance and upkeep, repairs are inadequate and the systems have been ineffective in raising productivity.

In this area, villages and village leagues can profit from Japan's experience in establishing mechanisms for the regulation of water distribution and in group monitoring of joint activities. Local hydraulic planners emerged who had superior powers of persuasion and an ability to negotiate extremely effectively with bureaucrats. These local leaders were not necessarily rich peasants. Neither, of course, were they extremely poor peasants. They are impossible to characterize in class terms based on the size of their landholdings. There have been great changes in the past century in role, scope of activity, and social background of local rural leaders, but there has been a strong tendency for the same villages to continue producing leaders of ability while others consistently fail to do so.

Thus we may conclude that it is the national corps of engineers that is the primary resource of technological development, that alone can combine all the structural factors (the five Ms) required, that alone can bridge old and new and foreign and domestic technologies, and that alone can develop technological fields that make possible a convergence of public and private and light and heavy industrial technologies. A policy for technological autonomy is one that includes the training of engineers and that allows room for engineers to bring out the needed latent links in the nation's technology.

Formation of Native Engineers

The role played by the rural leaders in agricultural technology was assumed by Japan's engineers in the factory. Despite the differences between agriculture and industry, we can speak of a Japanese type of engineer in both. This engineer made the necessary connections between traditional and modern technologies in each case. Engineers performed two roles in introducing basically new and unknown industrial technologies. As work leaders at the point of production, they (1) provided needed assistance to skilled workers and (2) trained workers to create a body of skilled workers. This work represented an outlook of engineering rooted in Japanese society and culture. The workers respected these leaders' practical ability to solve problems and followed their lead in the acquisition of skills.

The basis of this evaluation by the workers was not the educational background of the engineers. In Japan, in general, engineers are expected first to achieve mastery in a special field, but then also to gain experience in related fields and thus help to raise the level of national engineering expertise. Thus there is no narrow division of engineering fields based on special areas of expertise, such as between design and operations engineers. Instead, engineers take turns working in both areas. High performance, safety, manageability, and durability are therefore built into designs from the start because designs are tested for these qualities by engineers who know what to look for.

This ability has paid some remarkable dividends recently as the technological revolution proceeds, creating new interpenetrations between preexisting fields.

In the developing countries, not only are there few engineers to begin with, there is what can only be called a legacy of colonialism, a tight and narrow specialization. There are many techno-scientists, but few engineers actually at work on the sites where they are needed. Here these countries have a problem.

It is important to raise the status and widen the scope of activity of practical engineers, but above all, their numbers must be increased. However, this cannot be done simply by instituting the necessary secondary education. Study must follow practical experience and lead to further study; quality must be raised. Engineers' backgrounds must be broadened. Engineers must be produced who fit their nations, who can discover the necessary links discussed above and who can lead their nations in the process of development.

The discovery of latent links in development is the job of native engineers and creative managers, or at times of what might be termed engineer-managers.

To create a corps of engineers across a broad span of industrial activity, the publication of basic texts in engineering and technology is required. A great many other things that have not been done need to be done; for example, the establishment of correspondence courses and practical and licence testing.

There is an ongoing debate about whether technical and scientific education should take place in the language of the nation in which it is being administered or in a foreign language. Here our position is clear. For the dissemination of engineering knowledge and the development of latent links, it is necessary to educate in the local language. The techno-scientists must decide on the translation of technical terms that will best aid the spread of the desired engineering knowledge. That this is especially important in agricultural technology is obvious from the Japanese experience.

The agricultural, fishery, and industrial experiment stations and pilot farms the government established throughout Japan could meet specific needs in these industries, acting also as aids to the establishment of necessary links between pre-existing technologies and industries and the needs of modernization.

Manning these stations is not difficult to effect. Staff need not be university-trained experts; if they know the technology and are able to keep records properly and handle ordinary questions but know when to refer more complicated questions to experts, this is sufficient. This work, in turn, serves engineers and scientists in carrying out their work.

Public Management of Technology

Management techniques can be taught, but education alone cannot create technology management. In any age, there are few truly creative managers. We discussed the Japanese style of management above, so we will not elaborate on it here. As we noted in our discussion of the Japanese style of engineer formation, there have been different types of technology management in each period and sector. We must repeat, however, that management is a decisive, vital factor in technology formation and development.

We wish to lay stress on the importance of technology management rather than the technology of business management. Until now, this theme has been almost completely absent in our dialogue.

In machine technology, for example, there is no piece of machinery and no factory in which operation will proceed according to the manual from the very beginning. All engineers know that machinery tends to be operated, over the long run, at less than its optimum capacity. Speeding its operation up to normal capacity and making this normal is the job of the engineer, especially the chief engineer.

Links with related technologies and supporting services are essential, but the core of technology management is the establishment of stable qualitative and quantitative standards for raw materials, processing these precisely, adding parts and seeing that the product is turned out on time and according to standards.

One factory encouraged its workers to "work rationally, save costs, and maintain quality," and this seems right on the mark. Nothing else helps prevent breakdown more than safe, stable, and proper operation. Securing a stable supply of materials and energy is a basic principle of economy, and a good standardized product is a sure sign of high quality-control standards. We call this principle of technology management SERQ: safety, economy, rationality, and quality.

When engineers and managers put SERQ thoroughly to work, they often change the very layout of a factory. This results in safer, more productive, more quality-controlled conditions. The manufacturing process becomes highly systematized, which produces changes in the mode of operation. Overall production can be improved and the chance of error reduced through the use of electronic machinery, but the basic concept remains in the hands of the manager or the design engineer. SERQ must be applied creatively.

In Japan, there is a dual structure in both the economy and technology; that is, a very large-scale sector and a sizable small-scale business sector. Both are well integrated into the national economy and technology network; both have their own potentials for technology innovation, and the small sector is capable of producing and developing its own technology, suited to its needs.

The rational breakdown and division of the manufacturing process is driven forward by the two forces of the need for a new approach to managing factory operations and the desire of skilled workers for autonomy. The two combine to bring about a dual structure in the economy and in technology. This dual structure originated from the needs of labour, business, and technology management; it is a consequence of the prevailing economic and technological conditions.

One-time farmers found a new identity as skilled factory hands and were then able to set themselves up independently as small-business men. How was this possible? One answer is that the Japanese farmer (whether freeholder or tenant) was an agricultural manager. Self-management was firmly rooted in the peasantry, and highly skilled people have the confidence to deal with technology. These two factors doubtless combined to produce this phenomenon.

Just as machinery must be well taken care of lest its performance drop and its life span shrink, so technology on the whole must be adjusted and managed to suit each manufacturing process, each factory, and each related field or it will not perform effectively. In most manufacturing industries, an unstable supply of electric power or other related services produces ill effects that increasingly worsen as the modern technology becomes more highly sophisticated and larger in scale. If the energy supply is unstable, industries are forced to produce and regulate their own electric power at high cost. Such costs can be avoided if there is adequate public supervision and maintenance of technology and its related services.

Energy control systems and related services should be established on a national level. This, in turn, would raise SERQ standards in each company and factory, and less investment would mean a more rapid division in the manufacturing processes.

Social stability and public welfare are better assured if technology management is operating effectively on a national scale. At the same time, this will act as a preventative against severe pollution problems.

Notes

1. Among the technologies rated high internationally at the time of this project were semiconductor manufacturing; optoelectronics, including optical fibres; composite materials, such as carbon fibres; ceramics; and antiheat materials. Some areas of biotechnology and fermentology were also rated high. Since 1985, the superconductor has come to attract much attention world-wide.
2. Mark Gayn, *Japan Diary* (Tokyo and Rutland, Vt., Tuttle, 1981), p. 1.
3. Arisawa Hiromi et al. (1967), vol. 2, p. 6. Much of the following discussion has relied on this source and on its companion volume, *Shōwa keizai shi* (History of the Japanese economy in the Showa period) (1980).
4. Regarding the "development" effect of Malayawata Steel, a cross-industry analysis by Professor Torii Yasuhiko of Keio University, Tokyo, is available in the October 1978 issue of *Tekkokai*. Our joint comparative studies of the Japanese steel-manufacturing technology transfers to Malaysia and Brazil remain to be published.
5. Hoshino Yoshiro, *Mohaya gijutsu nashi* (No more new technology) (Tokyo, Kappa Books, 1978), p. 15.
6. A chip consisting of fewer than 100 elements is usually called an SSI, one with more than 100 but fewer than 1,000 an MSI, and one with more than 1,000 an LSI. A chip consisting of more than 10,000 elements is called a super LSI or, more commonly, a VLSI.
7. No one can guarantee, however, that there will be no problem in the future. For some time to come, to be sure, an increase in employment in (or a movement of the labour force to) the tertiary sector is considered probable, but there are also views that an economy with an annual growth rate of less than 5 per cent will give rise to unemployment. Yet, one way technological innovation might be justified concerns the fact that Japan is now quickly becoming an ageing society. If the social burden to be borne by the younger generation is to be lessened, productivity must be increased.
8. The air pollution regulations in most European countries are less strict than in Japan, but as the problem of acid rain becomes more serious internationally, controls are being tightened.
9. Ministry of Agriculture and Commerce, ed., *Shokkō jijō* (The state of workers) (Tokyo, Government Printing Office, 1903).

252

10. The capability of automated machines is greater today than 10 years ago. This is especially true of sensors, which help enhance the precision of programming. But this does not mean that computerization and higher sensor efficiency compare favourably with the ability of a skilled worker, nor is there any reason in theory (technological or economic) why they should. Computerization may enhance a low-level skill and in some cases even replace it, but it cannot compare with high-level skills. Ironically, today's progress in micro-electronics may lead to a situation in which high-level skills will actually be prevented from being formed.

11. Tokyo and Kyoto imperial universities played key roles in meeting the needs of the state. In the early 1900s, the ratio of graduates of the imperial universities choosing positions in government service over the private sector was 55 to 45; in the 1920s, however, 3,238 of a total 5,025 graduates in one particular year went into the private sector, overturning the ratio to 65 to 35. The five major *zaibatsu* and the South Manchuria Railway Co., with its mining and iron-manufacturing departments, took the lion's share of the graduates.

According to Uchida Hoshimi, in the 1910s, the Japanese government began, for the first time, to employ scientific and engineering personnel as officials to engage in "research and development." See Uchida Hoshimi, "Distribution of University-Graduate Engineers and Technologists in the 1920s," *Tokyo Keizai Daigaku Kaishi* (Bulletin of Tokyo University of Economics) 152 (September 1987).

In response to the growing needs of the industrial and business worlds, the economics department of Tokyo Imperial University was separated from the Faculty of Law as the Faculty of Economics in 1919. In 1920, Keio and Waseda received official recognition as the first privately run universities in Japan; six other private universities were established soon after. In the same year, a commercial high school was upgraded to a college. In 1929, the Tokyo School of Technology became a national college, renamed the Tokyo Institute of Technology. It has had a leading role as an educational institution in technology and engineering.

12. A vivid report of this appears in Kume Kunitake, ed., *Tokumei zenken taishi Bei-Ō kairan jikki* (Account of the visit of the ambassador extraordinary and plenipotentiary to Europe and the United States), published in 1872/1873. See also Marlene J. Maya, "The Western Education of Kume Kunitake, 1871–76," *Monumenta Nipponica* 28, no. 1 (1973).

13. In agricultural technologies, intermediation refers to bridging the gap between the expertise and knowledge found in experimental stations and actual farming practices; intermediation in technology in general is much more far-reaching than this simple example.

14. In a machine-based industry, differences in skills affect factory performance less than in an industry dependent on simpler tools and instruments. A tendency toward uniformity characterizes modern technologies, but the problem of skill remains, though the nature of the problem has changed.

15. There are some of the opinion that, in the area of atomic energy technology in Japan, two opposing categories of engineer, theorists and experimentalists, are embroiled in a sort of tribal antagonism toward each other. If so, I am unable to determine whether it is because Japanese atomic energy technology has not yet attained independence or because such large technologies are bound to make polarization unavoidable. I would prefer to tentatively support the first view.

16. Observations in 1987 at several steel mills in north-eastern China revealed that generally no master plan for renewing technologies in the mill is present. Instead,

we found, for example, that a process in need of modernization was completely replaced by a set of equipment from abroad, even though the operational capacity of the new set far exceeded manufacturing capacity. More than half the new set's potential has therefore been left untapped. This kind of diseconomy is sometimes unavoidable when there is a dependence on foreign technology, but it is puzzling to find no plans for future adjustment and a seeming lack of awareness on the part of the mill's management of their need.

According to Japanese practice, technology development is in the hands of the chief engineer. In China, however, the culture of technology differs from the Japanese approach, and a native Chinese solution has yet to be found.

And probably for the same reason, there are no operational standards in Chinese industries. This is certain to be a barrier to the development of industrial technologies. Although technological levels are sufficiently high to produce locomotives domestically, they must be sent back to their place of manufacture for maintenance and repairs. This is because, with a lack of standardized processes and systems of operation, the technology remains with veteran workers.

17. With the explosive technological innovation that we see today, it often happens that the economic life of equipment will end before its theoretical life; so, while it may be of no further use in one country, it may well be useful in another, and the wise entrepreneur will keep an eye out for this kind of opportunity. Modern equipment will not operate efficiently in countries that experience frequent power failures or voltage changes. Proper operation would require home generators, rectifiers, etc., pushing up costs and lowering international competitiveness.

18. For an idea of how the people of this social stratum lived at the end of the nineteenth century, see the writings of Higuchi Ichiyo, one of the most prominent female writers of the Meiji period, who wrote about the difficulties of the young women of her time and who herself suffered poverty and an early death. In English, see Robert Lyons Deuly, *In the Shade of Spring Leaves—The Life and Writings of Higuchi Ichiyo, a Woman of Letters in Meiji Japan* (New Haven, Yale University Press, 1982). Refer especially to "Child's Play," "On the Last Day of the Year," and "Troubled Water."

19. For a vivid description of the people of this stratum and their lives in Tokyo before industrialization, see Matsubara Iwagoro's *Saiankoku no Tōkyō* (*Darkest Tokyo*) (Tokyo, Minyusha, 1893; reprint, Iwanami Shoten, 1988) (The English version is by F. Schroeder, *In Darkest Tokyo: Sketches of Humble Life in the Capital of Japan* [Yokohama, 1898]).

20. Yokoyama Gennosuke, *Nihon no kasō shakai* (Tokyo, Kyobunkan, 1899; reprint, Iwanami Shoten, 1949).

21. Sawsan al-Messiri, *Ibn al-balad—A concept of Egyptian identity* (Leiden, 1978).

22. Refer to Akimoto Ritsuro, *Gendai toshi no kenryoku kōzō* (Power structure in the contemporary city) (Tokyo, Aoki Shuppan, 1971).

23. These handbooks were technical manuals that reflected "the results of Japanese study of agriculture influenced, not by modern, Western science, but by Chinese agricultural studies such as the sixth-century *Qi min yao shu*." See Furushima Toshio, *Nōsho no jidai* (The age of agriculture books) (Tokyo, Nosangyoson Bunka Kyokai, 1980).

24. Refer, for example, to Nishioka Toranosuke, *Kinsei ni okeru ichi rōnō no shōgai* (The life of an old farmer in the modern age) (former title, *Rōnō Watanabe Fushōō den* (The life of an old farmer, Fushoo) (Tokyo, Kodan Sha, 1978). The venerable Watanabe was born in 1793 in Akita Prefecture and died in 1856. He

was a contemporary of the famous agricultural reformer Ninomiya Sontoku (1787–1856). Many other agricultural reformers were active in this period.

25. See Iinuma Jiro, *Nihon nōgyō no sai-hakken—Rekishi to fūdo kara* (The rediscovery of Japanese agriculture—From its history and climate) (Tokyo, NHK Books, 1975).

26. See the survey report by Hatate Isao (1981).

27. For a fuller treatment of this and related questions, see Toyoda Toshio, ed., *Vocational Education in the Industrialization of Japan* (Tokyo, United Nations University, 1987).

28. There is an excellent study on foot-bellows iron making by Ohashi Shuji, *Bakumatsu Meiji seitetsu shi* (The history of iron manufacture at the end of the Tokugawa Era and Meiji) (Tokyo, Agunesu Sha, 1975). For an explanation of the manufacturing method, refer to the July 1984 issue of *Boisu* (Voice).

29. More precisely, the funds came through the Yokohama Specie Bank (now the Bank of Tokyo). This money was from individual savings deposits at post offices throughout the nation; these savings accounts constituted the government's most important source of financing for industrial investment.

30. Even today, in Kitakyushu, there is an organization for the recruitment of day-labourers.

31. During the Sino-Japanese War, the Chinese armed forces, even the navy, had much better weapons (though they lacked a rapid-fire gun), but they were defeated in large part because of problems in supply and maintenance. A common problem at the initial stage of industrialization is a neglect of maintenance (parts supply and repairs) and management skills.

32. See the recent study by Sugiyama Shinya, "Nihon sekitangyō no hatten to Ajia sekitan shijō" (Development of the Japanese mining industry and the Asian coal markets), *Gendai Keizai*, Spring 1982. See also the works by Kasuga Yutaka, one of our project participants: "Kan-ei Mi'ike Tankō to Mitsui Bussan" (The government-operated Mi'ike Coal-Mine and Mitsui & Co.), *Mitsui Bunko Ronso*, 1976, no. 10; "Mitsui Zaibatsu ni okeru sekitangyō no hatten kōzō" (Development structure of coal-mining in the Mitsui financial clique), ibid., no. 11; "1910-nen-dai ni okeru Mitsui Kōzan no tenkai" (The development of Mitsui mines in the 1910s), ibid., no. 12; and "Mitsui Tankō ni okeru 'gorika' no katei" (Process of 'rationalization' at the Mitsui Coal-Mine), ibid., no. 14.

33. On this system and the refining methods, see the works by Sasaki Junnosuke (1979, 1980).

34. Regarding the *tomoko*, see Murakushi Nisaburo's "Tomoko kenkyū no kaiko to kadai—Nihon kōfu kumiai kenkyū josetsu no isseki to shite" (Retrospectives and tasks in studies of the *tomoko*—As a part of the introduction to the study of the miners' unions in Japan), *Keizai Shirin* 48, no. 3 (1980). See also Matsushima Shizuo's *Tomoko no shakaigaku-teki kōsatsu* (A sociological study of the *tomoko*) (Tokyo, Ochanomizu Shobo, 1978).

35. Murakami Yasumasa and Hara Kazuhiko, *Gijutsu no shakai shi* (A social history of technology), vol. 4 (Tokyo, Yuhikaku, 1982), p. 51.

36. On the other hand, coal-mining requires a technology for preventing gas explosions in the pits that of course is not necessary for metal mines.

37. Murakami Yasumasa, Hara Kazuhiko (note 35 above). My description in this part is based on Hoshino Yoshiro, *Ashio Dōzan no gijutsu to keiei no rekishi* (*History of Technological and Administrative Development in the Ashio Copper Mine*) (Tokyo, United Nations University Press, 1982), HSDRJE-79J/UNUP-

403, and also Shoji Yoshiro and Sugai Masuro, *Tsūshi Ashio kōdoku jiken, 1877–1984* (A general history of poisoning incidents at Ashio Copper Mine) (Tokyo, Shin'yo Sha, 1984).

38. Inoue had gone to England in 1863 with Ito Hirobumi and others; he studied civil engineering and geology and returned to Japan in 1868, after which he joined the new government.

39. The Shinkansen was opened in 1964 to connect the 515 kilometres from Tokyo to Osaka in a travel time of only 3 hours, unprecedented in the world. The line was later extended to Kitakyushu, and two other routes have since been added. The train operates at a maximum speed of 210 kilometres per hour and an average speed of 170 kilometres per hour. The most advanced technology has been adopted in several areas of operation, notably in the chassis, structure, signal system, and safety controls.

40. It is important to note that the position of chairman of the Railway Council, the nucleus of railway policies, had been occupied by a vice-chief of the army general staff, and, moreover, two members of the council were also from the army. The private railway companies were obliged to co-operate with the military in times of an emergency. The military reserved the right to control nation-wide operation schedules and determine standards for major specifications for trains, stations, signalling systems, etc. In other words, in addition to the presence of economic and technological obstacles, there was the problem of the power of the military to shape railway policy to suit its particular needs.

41. See Takeuchi (1979) for a detailed account of this.

42. See Yonekawa Shin'ichi, "Bōsekigyō ni okeru kigyōseichō no kokusai hikaku" (International comparison of growth of enterprises in the textile industry), *Keizai Kenkyu* (Hitotsubashi University, Tokyo) (October 1978).

43. See Izumi (1979a).

44. See Yamamoto Shigemi, *Aa Nomugi Tōge* (Ah, Nomugi Pass) (Tokyo, Kadokawa Bunko, 1952); this work is based on interviews with over 300 elderly women who had once worked in the spinning mills. It was filmed and reportedly left a strong impression on viewers.

45. Ministry of International Trade and Industry, "Industrial Statistics: Industry" (Tokyo, GPO, 1980).

46. The surveys were conducted at the end of 1980 for Japan, in 1977 for the United States, and in 1980 for West Germany. In Japan, a medium-sized enterprise was defined as one with fewer than 300 employees; in the United States, fewer than 250; and in some European countries, fewer than 500.

47. The OEM system was adopted in the food industry before the development of the machine industry, although there it is referred to as "packer and brand owner," that is, processor and distributor system.

48. See Okumura (1973).

49. *Human Resources for Economic Development* (Geneva, ILO, 1966), p. 5.

50. Japanese students studying in England were advised by their professors there to transfer to German institutions to better understand the process of catching up. An intriguing twist to this was provided by Henry Dyer (1848–1918), a graduate of Glasgow University, who was invited to Japan to be the principal instructor in the College of Engineering of the Ministry of Industry, which had been established under the recommendation of Edmond Morell. Dyer returned to Britain, and, from his experiences training engineers in Japan, he promoted a major reform of engineering education in Britain. For more on the Dyer experiment,

refer to Kita Masami of Soka University, "Kōbu Daigakkō to Gurasugo Daiga-ku" (The College of Works and Glasgow University) (*Shakai Keizai Shigaku* 46, no. 5).

51. See Sumiya Mikio, *Nihon shokugyō kunren hattatsu shi* (A history of the development of Japanese vocational training), 2 vols. (Tokyo, Nihon Rodo Kyokai, 1952) 1:216 ff.

52. Nagai Michio, *Kindaika to kyōiku* (Modernization and education) (Tokyo, University of Tokyo Press, 1969) p. 102 ff.

53. Morikawa Hidemasa, *Gijutsu-sha—Nihon kindai-ka no ninaite* (Engineers—The modernizers of Japan) (Tokyo, Nihon Keizai Shimbun Sha, 1975).

54. Miyoshi Shinji, *Meiji no enjinia kyōiku: Nihon to Igirisu no chigai* (Engineering education in the Meiji period: Differences between Japan and England) (Tokyo, Chuo Koron Sha, 1983).

55. On Tokyo Imperial University, see Nakayama Shigeru, *Teikoku daigaku no tanjō* (Birth of the imperial university) (Tokyo, Chuo Koron Sha, 1978).

56. Regarding this dictionary, see the recent work by Takada Hiroshi, *Kotoba no umi e* (Towards the sea of words) (Tokyo, Shincho Sha, 1978).

57. Hara Akira, "Zaikai" (Business circle), in *Kindai Nihon kenkyū nyūmon* (An introduction to the study of modern Japan) (Tokyo, University of Tokyo Press, 1977).

58. Sakaguchi Akira, "Zaikai, seitō, kanryō" (Financial circles, political parties, and bureaucrats), in vol. 4 of *Nihon no kigyō to kokka* (Japanese enterprises and the state), ed. Morikawa Hidemasa (Tokyo, Nihon Keizai Shimbun Sha, 1976).

59. Shishido Toshio, *Mitsubishi shōji no kenkyū* (A study of the Mitsubishi Trading Company) (Tokyo, Toyo Keizai Shimpo Sha, 1970), pp. 194–95.

60. The institute's formal name was the Institute of Physics and Chemical Research. This institute introduced new policies in organization and the administration of research that are still of use and recognized effectiveness today.

61. Yokoyama Gennosuke, *Naichi zakkyo-go no Nihon* (Tokyo, Iwanami Shoten, 1954). See his preface.

62. Arisawa Hiromi et al., eds., *Nihon sangyō hyakunen shi* (A 100-year history of Japanese industry), 2 vols. (Tokyo, Nihon Keizai Shimbun Sha, 1967), p. 141; Goto Shin'ichi, *Nihon kin'yū seido hattatsu shi* (History of the development of the Japanese financial system) (Tokyo, Kyoiku Sha, 1980), p. 11.

63. See Kato Toshihiko, *Hompō ginkō shi ron* (History of Japanese banks) (Tokyo, University of Tokyo Press, 1957).

64. See *Meiji zaiseishi* (History of Meiji financial policies), vol. 13 (Tokyo, Government Printing Office).

65. The statement of purpose is from the government's official policy at the time.

66. Katsura Yoshio, *Sōgō shōsha no genryū: Suzuki Shōten* (Suzuki Shoten: Origin of a general trading company) (Tokyo, Nihon Keizai Shimbun Sha, 1977).

67. Nissan Konzern, the predecessor of present-day Nissan Motors, was a "new *zaibatsu*," in that it was not a family-owned, family-run business. Furthermore, unlike such *zaibatsu* as Mitsui, Mitsubishi, and Sumitomo, Nissan did not have its own bank, and, as a result, after the war, it suffered more from the dissolution of the *zaibatsu* than the others. Ayukawa Gisuke, the founder of Nissan, was a leading figure in the new heavy-industry complex in Japan. He was an engineer originally in the casting industry, finally diversifying into fishing and automobiles. In order to compete with the other *zaibatsu*, he established manufacturing in Manchuria. Ayukawa was of a new breed, but, his dependence on Manchuria and

his lack of a main bank for financial support eventually hurt him. The old *zaibat-su*, on the other hand, diversified their businesses, but only into areas or industries with which their existing businesses had some relation; in other words, they merely extended their existing activities.

Mitsui, for example, was careful to avoid any risks that could threaten the property of the Mitsui family.

68. Mishima Yasuo, "Mitsubishi Shōji—Zaibatsu-gata Shōsha no Keisei" (The Mitsubishi Corporation—The formation of a *zaibatsu*-type trading company), *Keieishigaku* 8, no. 9. See also Noda Kazuo et al., eds., *Kindai Nihon Keieishi no Kiso Chishiki* (Basic knowledge of modern Japanese management history) (Tokyo, Yuhikaku, 1981).

69. *Fortune* 102(3): 188–206.

70. These figures and those in the following pertaining to Mitsubishi are from the company's annual report of 1981.

71. The figures pertaining to Mitsui and Sumitomo corporations are from their respective annual reports of 1981.

72. Nakagawa Keiiehiro, "Nihon no Kōgyōka Katei ni Okeru 'Soshiki Sareta Kigyōsha Katsudō'" (Organized entrepreneurial activities in the process of Japanese industrialization), *Keieishigaku* 2, no. 3.

73. Mishima Yasuo, "Sōgō shōsha—Sengo ni okeru kenkyū-shi" (General trading firms—A history of studies after the war), in vol. 3 of *Nihon keieishi kōza* (Tokyo, Nihon Keizai Shimbun Sha, 1976).

74. Ibid.

75. Nikko Research Center, ed., *Mitsubishi Shōji no kenkyū* (A study of Mitsubishi Corporation) (Tokyo, Toyo Keizai Shimpo Sha, 1980), p. 9.

76. See Mishima (1976).

77. Nikko Research Center, *Mitsubishi Shōji*, p. 23.

78. Misonoi Hitoshi, "Sōgō shōsha wa shayō dearu ka" (Is the general trading firm declining?), *Ekonomisto*, 28 May 1961.

79. Nikko Research Center, *Mitsubishi Shōji*, pp. 18–19.

80. Sato Seizaburo, "Meiji Ishin no saikentō" (Re-examination of the Meiji Restoration), in *Kindai Nihon kenkyū nyūmon* (An introduction to the study of modern Japan) (Tokyo, University of Tokyo Press, 1977).

81. This was published by the Ministry of Agriculture and Commerce in 1884.

82. There is an English translation of the work: *Tien-Kung K'ai-Wu: Chinese Technology in the Seventeenth Century*, by E-tu Zen Sun and Shiou-Chuan Sun (Pennsylvania State University Press, 1965). What is interesting is that this work (extant in Japan in various editions and copied versions) was not widely read in China and had even been lost at one time. It was brought back to China by returning scholars, and attracted a great deal of interest (Yabu'uchi 1969: 3).

83. See Yuasa (1984), vol. 2, p. 491.

84. See, for example, Hiroshige Toru, *Kagaku no shakai-shi—Kindai Nihon no kagaku taisei* (A social history of science—The structure of modern Japanese science) (Tokyo, Chuo Koron Sha, 1960); idem, *Kagaku to rekishi* (Science and history) (Tokyo, Misuzu Shobo, 1965); idem, *Kindai kagaku saikō* (Modern science reconsidered) (Tokyo, Asahi Shimbun Sha, 1979); Nakayama Shigeru, *Kagaku to shakai no gendai-shi* (Contemporary history of science and society) (Tokyo, Iwanami Shoten, 1981); Hiroshige Toru, Ito Shuntaro, and Murakami Yoichiro, *Shisō-shi no naka no kagaku* (Science in the history of thought) (Tokyo, Mokutaku Sha, 1975). This last work provides a concise summary of issues relating to the history of science.

85. See Okakura Koshiro, "Unesco kankoku to Kagakusha Kenshō" (Unesco recommendations and the Scientists' Charter), in Watanabe Naotsune and Igasaki Akio, eds., *Kagakusha Kenshō* (Scientists' Charter) (Tokyo, Keiso Shobo, 1980), p. 29; Yuasa Mitsutomo, *Nihon no kagaku-gijutsu hyakunen-shi* (A 100-year history of Japanese science and technology), 2 vols. (Tokyo, Chuo Koron Sha, 1984).

86. *Kagaku gijutsu no shi-teki tenkai* (A historical development of science and technology) (Tokyo, Government Printing Office, 1980), p. 42.

87. Murakami Nobuhiko, *Meiji josei-shi* (History of women in the Meiji period) (Tokyo, Kodan Sha, 1977), vol. 1, p. 53.

88. See Yokoyama (1949), pp. 236–38.

89. Ibid., p. 239.

90. Ibid.

91. Ibid., p.241.

92. Nakayama Ichiro, ed., *Chingin kihon chōsa* (Basic survey of wages) (Tokyo, Toyo Keizai Shimpo Sha, 1956).

93. See J. P. Aron's chapter on women spinners in his *La femme du XIXᵉ siècle* (Paris, 1980). There is a Japanese translation by Kataoka Yoshihiko et al., *Tsuyumichiura no joseishi* (Tokyo, Shinhyoron, 1984).

94. Shoji Yoko, "Home and Living," *Fujin hakusho* (White paper on women) (Tokyo, Sodo Sha, 1984), p. 51.

95. Shinozuka Eiko, *Nihon no joshi rōdō* (Female labour in Japan) (Tokyo, Toyo Keizai Shimpo Sha, 1982), p. 34.

96. For a fuller understanding of the problem provisionally addressed above, the reader may wish to refer to the following works (see Nakamura [1985]):
   – Kase Kazutoshi: "Nō-gyogyō ni okeru gijutsu kakushin to joshi rōdō" (Female labour and technological improvements in the agricultural and fishing industries)
   – Miyake Akimasa: "Toshi kasō no fujo rōdō" (The urban lower strata and female labour)
   – Nakamura Masanori: "Seishigyō ni okeru rōdōryoku no kōsei to rōshi kankei" (Labour relations and the composition of the work-force in the silk thread industry)
   – Nishinarita Yutaka: "Kōzan no gijutsu kakushin to joshi rōdō" (Female labour and improvements in mining technology)
   – Shiota Fukiko: "Sengo Nihon no gijutsu kakushin to joshi kōyō rōdō" (Female employment and technological improvements in post-war Japan)

97. Shukan-Asahi, ed., *Nedan no Meiji-Taishō-Shōwa fūzoku-shi* (A history of manners and customs in the Meiji, Taisho, and Showa periods, with special reference to prices) (Tokyo, Asahi Shimbun Sha, 1981).

98. *Watashi no jokō aishi* was written by Takai Toshio and published by Sodo Sha, Tokyo. Hayashi's work was first published by Chuo Koron Sha, Tokyo, in 1930. The latest edition is from Shincho Sha, Tokyo, 1979.

99. From her preface to *Ten no yamu* (Heaven is ill), ed. Ishimure Michiko (Tokyo, Asahi Shobo, 1974). See also her works *Kukai jōdo* (Bitter sea and paradise) (Tokyo, Kodan Sha, 1972) and *Tsubaki no umi no ki* (Camellia sea diary) (Tokyo, Asahi Shimbun Sha, 1980).

100. Harada Masazumi, *Minamata byō ni manabu tabi—Minamata byō no mae ni Minamata byō wa nakatta* (Journey in search of Minamata disease—There was no Minamata disease before its first occurrence) (Tokyo, Nihon Hyoron Sha, 1985), pp. 66–67, 144.

101. Shoji Yoshiro and Sugai Masuro, *Tsūshi Ashio kōdoku jiken, 1877–1984* (General history of pollution incidents at the Ashio Copper Mine, 1877–1984) (Tokyo, Shin'yo Sha, 1984).

102. The leader in the anti-pollution struggles was Tanaka Shozo (1841–1913), who fought for many years to bring about strict pollution control measures. He was defeated in his campaign, but regained recognition in the 1970s, at the time of renewed anti-pollution protests. See *Tanaka Shōzō zenshū* (The complete works of Tanaka Shozo), 18 volumes (Tokyo, Iwanami Shoten, 1977–80); see also, Kenneth Strong, *Ox Against the Storm: A Biography of Tanaka Shozo, Japan's Conservationist Pioneer* (Kent, Paul Norbury, 1977).

103. See Iijima Nobuko, *Kankyō mondai to higaisha undō* (Environmental problems and victims' protests) (Tokyo, Gakuyu Sha, 1984). See also her *Pollution Japan: Historical Chronology* (New York, Pergamon, 1980).

References

Abe Takeshi. 1983. "Meiji zenki ni okeru Nihon no zairai sangyō" (Japanese native industry in early Meiji). In *Matsukata zaisei to shokusan kōgyō seisaku*. *See* Umemura and Nakamura 1983.

Aoki Eiichi. 1979. *Chiiki shakai kara mita tetsudō kensetsu* (*Railway Construction as Viewed from Local Society*, HSDRJE-13/UNUP-97, 1980). UNU Research Paper No. HSDRJE-13J/UNUP-33. Tokyo, United Nations University Press.

———. 1982a. *Keiben tetsudō no hattatsu* (Light railway development—Networks and rolling stock). UNU Research Paper No. HSDRJE-50J/UNUP-355. Tokyo, United Nations University Press.

———. 1982b. *Toshika no shinten to tetsudō gijutsu no dōnyū* (The development of railways in Japan's big cities). UNU Research Paper No. HSDRJE-54J/UNUP-359. Tokyo, United Nations University Press.

Arisawa Hiromi et al., eds. 1967. *Nihon sangyō hyakunen shi* (A 100-year history of Japanese industry). 2 vols. Tokyo, Nihon Keizai Shimbun Sha.

———, eds. 1980. *Shōwa keizai shi* (History of the Japanese economy in the Showa period). 2 vols. Tokyo, Nihon Keizai Shimbun Sha.

Asai Yoshio. 1983. "Gaishi dōnyū to Nihon Kōgyō Ginkō" (Induction of foreign capital and the Industrial Bank of Japan). Tokyo, United Nations University. Unpublished research paper.

Asajima Shoichi. 1983. *Mujingyō no sonritsu kiban to sono henshitsu* (Japan's mutual loan business and how it changed). UNU Research Paper No. HSDRJE-87J/UNUP-488. Tokyo, United Nations University Press.

Carson, Rachel. 1962. *Silent Spring*. New York, Houghton Mifflin.

Chiba Osamu. 1983. "Nōgyō kin'yū ni okeru kyōdō kumiai" (Co-operative associations in agricultural financing). Tokyo, United Nations University. Unpublished research paper.

Chimoto Akiko. 1982. *Meiji shoki bōsekigyō no rōmu kanri no keisei* (The formation of labour control in the early Meiji cotton mills). UNU Research Paper No. HSDRJE-75J/UNUP-399. Tokyo, United Nations University Press.

Cho Yukio. 1981. *Ishoku-gata daikōgyō to zairai sangyō* (Large-scale transplanted industries and traditional industries: A comparison). UNU Research Paper No. HSDRJE-65J/UNUP-377. Tokyo, United Nations University Press.

261

Fujimori Terunobu. 1982. *Meiji no Tōkyō keikaku* (Tokyo plans in the Meiji period). Tokyo, Iwanami Shoten.

Fujita Teiichiro. 1981. *Kindai Nihon dōgyō kumiai shiron josetsu* (Introduction to a history of industrial associations in modern Japan). UNU Research Paper No. HSDRJE-64J/UNUP-376. Tokyo, United Nations University Press.

Furushima Toshio. 1980. *Nōsho no jidai* (The age of agriculture books). Tokyo, Nosangyoson Bunka Kyokai.

Gayn, Mark. 1981. *Japan Diary*. Tokyo and Rutland, Vt., Tuttle.

Goto Shin'ichi. 1980. *Nihon kin'yū seido hattatsu shi* (History of the development of the Japanese financial system). Tokyo, Kyoiku Sha.

Hara Akira. 1977. "Zaikai." In *Kindai Nihon kenkyū nyūmon* (An introduction to the study of modern Japan). Tokyo, University of Tokyo Press.

Harada Katsumasa. 1979. *Tetsudō dōnyū to gijutsu jiritsu e no tembō* (*Japan's Discovery, Import, and Technical Mastery of Railways*, HSDRJE-12/UNUP-51, 1979). UNU Research Paper No. HSDRJE-12J/UNUP-32. Tokyo, United Nations University Press.

―――. 1980. *Tetsudō gijutsu no jiritsu to kikakuka no shinkō* (*Technological Independence and Progress of Standardization in the Japanese Railways*, HSDRJE-36/UNUP-223, 1981). UNU Research Paper No. HSDRJE-36J/UNUP-209. Tokyo, United Nations University Press.

Harada Masazumi. 1972. *Minamata byō* (Minamata disease). Tokyo, Iwanami Shoten.

―――. 1985. *Minamata byō ni manabu tabi―Minamata byō no mae ni Minamata byō wa nakatta* (Journey in search of Minamata disease―There was no Minamata disease before its first occurrence). Tokyo, Nihon Hyoron Sha.

Hashimoto Tetsuya. 1980. *Chihō toshi no kasō minshū to minshū bōdō* (*The Lower Socio-economic Classes and Mass Riots in a Provincial City*, HSDRJE-32/UNUP-256). UNU Research Paper No. HSDRJE-32J/UNUP-205. Tokyo, United Nations University Press.

Hatate Isao. 1979. *Suiri to shidōshatachi* (*Irrigation Water Rights Disputes in Japan―as Seen in the Azusa River System*, HSDRJE-5/UNUP-87, 1979). UNU Research Paper No. HSDRJE-5J/UNUP-25. Tokyo, United Nations University Press.

―――. 1980. *Ogogawa, Yamadagawa sosui no seiritsu katei* (*The Establishing Process of the Ogo and Yamada Canals*, HSDRJE-44/UNUP-336, 1981). UNU Research Paper No. HSDRJE-44J/UNUP-217. Tokyo, United Nations University Press.

―――. 1981. *Nihon shihonshugi no seisei to fudō sangyō* (*Urbanization and the Real-Estate Business*, HSDRJE-57/UNUP-410, 1983). UNU Research Paper No. HSDRJE-57J/UNUP-369. Tokyo, United Nations University Press.

Hayashi Takeshi. 1979. *Gijutsu no iten, hen'yō, kaihatsu* (*Historical Background of Technology Transfer, Transformation, and Development in Japan*, HSDRJE-19/UNUP-48, 1979). UNU Research Paper No. HSDRJE-19J/UNUP-49. Tokyo, United Nations University Press.

Hiroshige Toru. 1960. *Kagaku to shakai-shi―Kindai Nihon no kagaku taisei* (A social history of science―The structure of modern Japanese science). Tokyo, Chuo Koron Sha.

―――. 1979. *Kagaku to rekishi* (Science and history). Tokyo, Misuzu Shobo.

Horii Kenzo. 1979. *Hata seki ni okeru suiri kōzō* (Organization of the irrigation system of the Hata Canal). UNU Research Paper No. HSDRJE-21J/UNUP-70. Tokyo, United Nations University Press.

Hoshino Yoshiro. 1979. *Hoshino Yoshiro chosakushū* (The collected works of Hoshino Yoshiro). 8 vols. Tokyo, Keiso Shobo.

―――. 1982. *Ashio Dōzan no gijutsu to keiei no rekishi* (*History of Technological and*

Administrative Development in the Ashio Copper Mine, HSDRJE-79/UNUP-417, 1983). UNU Research Paper No. HSDRJE-79J/UNUP-403. Tokyo, United Nations University Press.

————. 1986. *Sentan gijutsu no kompon mondai* (Basic problems of the latest technologies). Tokyo, Keiso Shobo.

Hosoi Wakizo. [1925] 1954. *Jokō aishi* (The sad story of the girl spinners). Tokyo, Iwanami Shoten.

Iida Ken'ichi. 1979. *Nihon tekkō gijutsu no keisei to tenkai (Origin and Development of Iron and Steel Technology in Japan*, HSDRJE-8/UNUP-89, 1980). UNU Research Paper No. HSDRJE-8J/UNUP-28. Tokyo, United Nations University Press.

————. 1981. *Kindai tekkō gijutsu no hatten to rōdōryoku* (Skilled labour and the development of steel technology in modern Japan). UNU Research Paper No. HSDRJE-53J/UNUP-358. Tokyo, United Nations University Press.

Iida Ken'ichi and Saegusa Hiroo. 1957. *Nihon kindai seitetsu-gijutsu hattatsu shi* (Development of modern iron- and steel-making technology in Japan). Tokyo, Toyo Keizai Shimpo Sha.

Iijima Nobuko. 1982. *Ashio Dōzan Yamamoto ni okeru kōgai* (Mining pollution in the Ashio Copper Mine). UNU Research Paper No. HSDRJE-76J/UNUP-400. Tokyo, United Nations University Press.

————. 1984. *Kankyō mondai to higaisha undō* (Environmental problems and victims' protests). Tokyo, Gakuyu Sha.

Iinuma Jiro. 1987. 2nd ed. *Nōgyō kakumei ron* (Agricultural revolution). Tokyo, Mirai Sha.

Ikeda Shoji, ed. 1982. *Niigata ken no kinzoku kakō sangyō* (Metal processing in industrial towns: The experience of Tsubame and Sanjo). UNU Research Paper No. HSDRJE-51J/UNUP-356. Tokyo, United Nations University Press.

Imamura Naraomi. 1979. *Tochi kairyō tōshi to nōgyō keiei (Land Improvement Investment and Agricultural Enterprises in Japan—as Seen in the Azusa River System*, HSDRJE-6/UNUP-88, 1980). UNU Research Paper No. HSDRJE-6J/UNUP-26. Tokyo, United Nations University Press.

————. 1980. *Toshika to chiiki nōgyō no tembō* (The progress of urbanization and prospects for regional agriculture). UNU Research Paper No. HSDRJE-42J/UNUP-215. Tokyo, United Nations University Press.

Industrial Bank of Japan, Research Department, ed. 1984. *Nihon sangyō dokuhon* (A reader of Japanese industry). Tokyo, Toyo Keizai Shimpo Sha.

Inoki Takenori. 1983. "Chiso beinō ron to zaisei seiri" (Financial settlement and the land tax in kind). In *Matsukata zaisei to shokusan kōgyō seisaku. See* Umemura and Nakamura 1983.

Ishihara Osamu. 1913. "*Eiseigaku-jō yori mitaru jokō no genkyō to kekkaku*" (Female workers and the sanitary conditions of the work-place). Tokyo, Kokka Igaku Kai.

Ishii Ichiro. 1979. *Nihon ni okeru dōro gijutsu no hattatsu* (The development of road construction technology in Japan). UNU Research Paper No. HSDRJE-10J/UNUP-30. Tokyo, United Nations University Press.

————. 1980. *Nihon ni okeru dōro gijutsu no hattatsu II* (The development of road construction technology in Japan—Part II). UNU Research Paper No. HSDRJE-35J/UNUP-208. Tokyo, United Nations University Press.

Ishii Tadashi. 1986. "Sen'i kikai gijutsu no hatten katei" (Development process of textile machinery). In *Kindai Nihon no gijutsu to gijutsu seisaku. See* Uchida, Nakaoka, and Ishii 1986.

Ishimure Michiko. 1972. *Kukai jōdo* (Bitter sea and paradise). Tokyo, Kodan Sha.

Ishizuka Hiromichi. 1977. *Tōkyō no shakai-keizai shi* (Socio-economic history of Tokyo). Tokyo, Kinokuniya Shoten.

———. 1979. *"Tōkyō-shi" kenkyū no hōhōron josetsu (Methodological Introduction to the History of the City of Tokyo,* HSDRJE-2/UNUP-85, 1981). UNU Research Paper No. HSDRJE-2J/UNUP-22. Tokyo, United Nations University Press.

———. 1980. *Toshi kasō shakai to "saimin" jūkyoron* (Lower socio-economic class urban society in Tokyo and housing for the poor—Up to the early 1920s). UNU Research Paper No. HSDRJE-28J/UNUP-201. Tokyo, United Nations University Press.

———. 1981. *Tōkyō no toshi suramu to kōshū eisei mondai (The Early History of the Control of Water-Borne Diseases in Tokyo,* HSDRJE-55/UNUP-409, 1983). UNU Research Paper No. HSDRJE-55J/UNUP-367. Tokyo, United Nations University Press.

Iwashita Masahiro. 1982. *Wasō orimono sangyō ni okeru zairai gijutsu to gairai gijutsu* (Traditional and transplanted technology in Japan's traditional clothing industry). UNU Research Paper No. HSDRJE-73J/UNUP-397. Tokyo, United Nations University Press.

Izumi Takeo. 1979a. "The Cotton Industry." *The Developing Economies* (Institute of Developing Economies, Tokyo). vol. 17, no. 4, 398–420.

———. 1979b. *Mengyō ni okeru gijutsu no hen'yō to kaihatsu (Transformation and Development of Technology in the Japanese Cotton Industry,* HSDRJE-25/UNUP-91, 1980). UNU Research Paper No. HSDRJE-25J/UNUP-74. Tokyo, United Nations University Press.

Jinnouchi Yoshito. 1981. *Denki kangai jigyō to chiiki shakai (Electric Irrigation and Local Society,* HSDRJE-58/UNUP-411, 1983). UNU Research Paper No. HSDRJE-58J/UNUP-370. Tokyo, United Nations University Press.

Kasuga Yutaka. 1982. *Hokkaidō sekitangyō no gijutsu rōdō (Transfer and Development of Coal Mine Technology in Hokkaido,* HSDRJE-48/UNUP-335, 1982). UNU Research Paper No. HSDRJE-48J/UNUP-262. Tokyo, United Nations University Press.

Kato Kozaburo. 1979. *Mengyō ni okeru gijutsu iten to keitai* (Technology transfer in the Japanese cotton industry). UNU Research Paper No. HSDRJE-18J/UNUP-68. Tokyo, United Nations University Press.

Kato Toshihiko. 1957. *Hompō ginkō shi ron* (History of Japanese banks). Tokyo, University of Tokyo Press.

Katsura Yoshio. 1977. *Sōgō shōsha no genryū: Suzuki Shōten* (Suzuki Shoten: Origin of a general trading company). Tokyo, Nihon Keizai Shimbun Sha.

Kawabe Nobuo. 1990. "Overseas Operations of the General Trading Company." In *General Trading Companies: A Comparative and Historical Study,* ed. Shin'ichi Yonekawa. Tokyo, United Nations University Press. Forthcoming.

Kikuura Shigeo. 1979. *Meiji shoki no garasu kōgyō no keifu* (History of the glass industry in the early Meiji era). UNU Research Paper No. HSDRJE-20J/UNUP-69. Tokyo, United Nations University Press.

Kiyokawa Yukihiko. 1980. *Sanhinshu no kairyō to fukyū dempa (The Development and Diffusion of Improved Hybrid Silkworms in Japan—The First Filial Generation,* HSDRJE-46/UNUP-253, 1981). UNU Research Paper No. HSDRJE-46J/UNUP-219. Tokyo, United Nations University Press.

Kobashi Ichiro. 1981. *Waga kuni ni okeru kaisha hōsei no keisei* (The formation of company law in Japan). UNU Research Paper No. HSDRJE-67J/UNUP-387. Tokyo, United Nations University Press.

Kobayashi Tatsuya. 1981. *Gijustsu iten—Rekishi kara no kōsatsu: Amerika to Nippon*

(Technology transfer—A study from the viewpoint of American and Japanese history). Tokyo, Bunshin Do.

———. 1983. *Zoku gijutsu iten—Dochakuka e no chōsen* (Technology transfer—Challenge to indigenization). Tokyo, Bunshin Do.

Kosuge Nobuhiko. 1980. *Waga kuni ni okeru jōsuidō no hattatsu* (*The Development of Waterworks in Japan*, HSDRJE-29/UNUP-240, 1981). UNU Research Paper No. HSDRJE-29J/UNUP-202. Tokyo, United Nations University Press.

Koyano Shogo. 1979. *Dentō sangyō gijutsu to shokunin no yakuwari* (*Technology of Traditional Industry and the Role of Craftsmen*, HSDRJE-1/UNUP-84, 1979). UNU Research Paper No. HSDRJE-1J/UNUP-21. Tokyo, United Nations University Press.

———. 1980. *Toshi kyojū ni okeru tekiō gijutsu no tenkai* (The development of technology adapted to urban life—Focusing on the lower socio-economic class). UNU Research Paper No. HSDRJE-27J/UNUP-200. Tokyo, United Nations University Press.

Kurasawa Susumu. 1982. *Kindai Nihon toshi keikaku kanren nempyō* (A history of city planning in Japan). UNU Research Paper No. HSDRJE-69J/UNUP-393. Tokyo, United Nations University Press.

Maeda Kazutoshi. 1990. "Business Activities of General Trading Companies." In *General Trading Companies: A Comparative and Historical Study*, ed. Shin'ichi Yonekawa. Tokyo, United Nations University Press. Forthcoming.

Masuda Hiromi. 1979. *Nihon ni okeru nairiku suiun no hattatsu* (Development of river transportation in Japan). UNU Research Paper No. HSDRJE-11J/UNUP-31. Tokyo, United Nations University Press.

———. 1980. *Shokusan kōgyō seisaku to Nobiruko* (*Japan's Industrial Development Policy and the Construction of the Nobiru Port: The Case Study of a Failure*, HSDRJE-37/UNUP-222, 1981). UNU Research Paper No. HSDRJE-37J/UNUP-210. Tokyo, United Nations University Press.

Matsubara Iwagoro. [1893] 1988. *Saiankoku no Tōkyō* (Darkest Tokyo). Tokyo, Iwanami Shoten.

Matsushima Shizuo. 1978. *Tomoko no shakaigaku-teki kōsatsu* (A sociological study of the *tomoko*). Tokyo, Ochanomizu Shobo.

Ministry of Agriculture and Commerce, ed. [1903] 1976. *Shokkō jijō* (Survey of workers' conditions). 3 vols. Tokyo, Shinkigen Sha.

Mishima Yasuo. 1976. "Sōgō shōsha—Sengo ni okeru kenkyūshi" (General trading firms—A history of studies after the war). In *Nihon keishi kōza* (History of Japanese business), ed. Nihon Keieishi Gakkai, vol. 3. Tokyo, Nihon Keizai Shimbun Sha.

MITI, Industrial Science and Technology Agency. 1981. *Sōzō-teki gijutsu rikkoku o mezashite* (Toward self-reliance in technology). Tokyo, Government Printing Office.

Miyoshi Shinji. 1983. *Meiji no enjinia kyōiku— Nihon to Igirisu no chigai* (Engineering education in the Meiji period: Differences between Japan and England). Tokyo, Chuo Koron Sha.

Morikawa Hidemasa. 1975. *Gijutsu-sha—Nihon kindai-ka no ninaite* (Engineers—The modernizers of Japan). Tokyo, Nihon Keizai Shimbun Sha.

———. 1976. *Nihon no kigyō to kokka* (Japanese enterprises and the state). Vol. 4 of *Nihon keieishi kōza* (History of Japanese business), ed. Nihon Keieishi Gakkai. Tokyo, Nihon Keizai Shimbun Sha.

Mori Kiyoshi. 1985. *Machi kōba kara no hassō* (Views from a small factory). Tokyo, Kodan Sha.

Mukae Yurio. 1982. *Yūbin chokin no hatten to sono shoyōin* (Development of post office savings schemes). UNU Research Paper No. HSDRJE-61J/UNUP-373. Tokyo, United Nations University Press.

Murakami Nobuhiko. 1973–81. *Meiji josei-shi* (History of women in the Meiji period). 3 vols. Tokyo, Kodan Sha.

Murakami Yasumasa, and Hara Kazuhiko. 1982. *Gijutsu no shakai shi* (A social history of technology). Vol. 1. Tokyo, Yuhikaku.

Murakushi Nisaburo. 1979. *Nihon sekitangyō no gijutsu to rōdō* (*Technology and Labour in Japanese Coal Mining*, HSDRJE-17/UNUP-82, 1980). UNU Research Paper No. HSDRJE-17J/UNUP-37. Tokyo, United Nations University Press.

———. 1981. *Manshū e no sekitangyō gijutsu iten to rōdōryoku* (*The Transfer of Coal-Mining Technology from Japan to Manchuria and Manpower Problems—Focusing on the Development of the Fushun Coal Mines*, HSDRJE-47/UNUP-225, 1981). UNU Research Paper No. HSDRJE-47J/UNUP-220. Tokyo, United Nations University Press.

Muroyama Yoshimasa. 1983. "Matsukata defurēshon no mekanizumu" (Mechanism of the Matsukata deflationary policy). In *Matsukata zaisei to shokusan kōgyō seisaku. See* Umemura and Nakamura 1983.

Nagai Michio. 1969. *Kindaika to kyōiku* (Modernization and education). Tokyo, University of Tokyo Press.

———, ed. 1984. *Development in the Non-Western World*. Tokyo, United Nations University Press.

Nagai Michio, and Miguel Urrutia, eds. 1985. *Meiji Ishin: Restoration and Revolution*. Tokyo, United Nations University Press.

Nagata Keijuro. 1980. *Tameike kangai chitai no nōgyō seisan to suiri* (Agricultural production and irrigation in areas dependent on reservoirs). UNU Research Paper No. HSDRJE-43J/UNUP-216. Tokyo, United Nations University Press.

Nakagawa Kiyoshi. 1982. *Senzen Tōkyō no toshi kasō* (The lower classes in prewar metropolitan Tokyo). UNU Research Paper No. HSDRJE-56J/UNUP-368. Tokyo, United Nations University Press.

Nakagome Shozo. 1982. *Ifuku sangyō no hajime* (The establishment of the Western clothing industry in Japan). UNU Research Paper No. HSDRJE-83J/UNUP-407. Tokyo, United Nations University Press.

Nakamura Hachiro. 1979. *Senzen no Tōkyō ni okeru chōnaikai* (*Town Organizations in Prewar Tokyo*, HSDRJE-3/UNUP-86, 1980). UNU Research Paper No. HSDRJE-3J/UNUP-23. Tokyo, United Nations University Press.

———. 1980. *Chōnaikai no soshiki to un'eijō no mondai ten* (The organization of town associations and their operational problems: A case-study of prewar Tokyo). UNU Research Paper No. HSDRJE-31J/UNUP-204. Tokyo, United Nations University Press.

———. 1982. *Denki jigyō no ranshō to tenkai katei* (The origin and development of electric power supply in Japan). UNU Research Paper No. HSDRJE-70J/UNUP-394. Tokyo, United Nations University Press.

Nakamura Hideichiro. 1982. *Chūshō kigyō keieisha no shisō* (Ideology of small enterprise management). UNU Research Paper No. HSDRJE-81J/UNUP-405. Tokyo, United Nations University Press.

Nakamura Masanori. 1952. "Kaisetsu" (Annotation). In *Aa Nomugi Tōge. See* Yamamoto 1952.

———. 1985. *Gijutsu kakushin to joshi rōdō* (*Technological Innovation and Female Labour in Japan*, forthcoming). Tokyo, United Nations University Press.

Nakamura Takafusa. 1978. *Nippon keizai—Sono kōzō to seichō* (The Japanese economy—Its structure and growth). Tokyo, University of Tokyo Press.

————. 1983. "Matsukata defurēshon zenshi" (Background of the Matsukata deflationary policy). In *Matsukata zaisei to shokusan kōgyō seisaku.* See Umemura and Nakamura 1983.

————. 1986. *Shōwa keizai-shi* (Economic history of the Showa period). Tokyo, Iwanami Shoten.

Nakaoka Tetsuro. 1980. *Kōjō no tetsugaku* (Philosophy of the factory). Tokyo, Heibon Sha.

————. 1986. "Gijutsushi no shiten kara mita Nihon no keiken" (The Japanese experience viewed from the history of technology). In *Kindai Nihon no gijutsu to gijutsu seisaku.* See Uchida, Nakaoka, and Ishii 1986.

Nakayama Ichiro. 1956. *Chingin kihon chōsa* (Basic survey of wages). Tokyo, Toyo Keizai Shimpo Sha.

Nakayama Shigeru. 1978. *Teikoku daigaku no tanjō* (Birth of the imperial university). Tokyo, Chuo Koron Sha.

————. 1981. *Kagaku to shakai no gendai-shi* (Contemporary history of science and society). Tokyo, Iwanami Shoten.

Nakura Bunji. 1980. *Nihon tekkōgyō to "nan'yō" tekkō shigen (The Prewar Japanese Steel Industry and Iron Ore Resources in South-east Asia: The Development of Malaysian Iron Ore by the Ishihara Sangyo Company,* HSDRJE-33/UNUP-235, 1981). UNU Research Paper No. HSDRJE-33J/UNUP-206. Tokyo, United Nations University Press.

Namie Ken. 1981. *Watashi no nōmin kyōiku no jissen (Farmer Education: A Personal Experience,* HSDRJE-62/UNUP-413, 1983). UNU Research Paper No. HSDRJE-62J/UNUP-374. Tokyo, United Nations University Press.

Namigata Shoichi. 1981. *Chōsen ni okeru kin'yū kumiai* (Development and function of co-operative financing associations in Korea). UNU Research Paper No. HSDRJE-60J/UNUP-372. Tokyo, United Nations University Press.

Nihon Keizai Shimbun Sha, ed. 1978. *Asu no raibaru—Oiageru Ajia no kikai kōgyō* (Tomorrow's rival—The Asian machine manufacturing industry in pursuit). Tokyo, Nihon Keizai Shimbun Sha.

Nishioka Toranosuke. 1978. *Kinsei ni okeru ichi rōnō no shōgai* (The life of an old farmer in the modern age). Tokyo, Kodan Sha.

Okakura Koshiro. 1980. "Unesco kankoku to Kagakusha Kenshō" (Unesco recommendations and the Scientists' Charter). In Watanabe Naotsune and Igasaki Akeo, eds., *Kagakusha Kenshō* (Scientists' Charter). Tokyo, Keiso Shobo.

Okumura Shoji. 1970. *Hinawa-jū kara kurofune made—Edo-jidai gijutsu shi* (Matchlock to black ships—History of technology in the Edo period). Tokyo, Iwanami Shoten.

————. 1973. *Koban, kiito, watetsu—Zoku Edo-jidai gijutsu shi II* (Kobang, raw silk, Tatara steel—History of technology in the Edo period, II). Tokyo, Iwanami Shoten.

Otsuka Tsutomu. 1982. *Shoku ni okeru Nihon no kindaika* (Transformation of Japan's eating habits). UNU Research Paper No. HSDRJE-82J/UNUP-406. Tokyo, United Nations University Press.

Saito Osamu. 1983. "Chihō reberu no shokusan kōgyō seisaku" (Industrial policies at the local level). In *Matsukata zaisei to shokusan kōgyō seisaku.* See Umemura and Nakamura 1983.

Saito Toshihiko. 1982. *Kinhonisei ka no zaigai seika* (Specie held abroad under the gold standard system). UNU Research Paper No. HSDRJE-68J/UNUP-388. Tokyo, United Nations University Press.

————. 1983. *Gaikoku kawase ginkō no seiritsu* (Foreign exchange banks in Japan and how they evolved). UNU Research Paper No. HSDRJE-88J/UNUP-489. Tokyo United Nations University Press.

Sakaguchi Akira. 1976. "Zaikai, seitō, kanryō" (Financial circles, political parties, and bureaucrats). In *Nihon no kigyō to kokka*. See Morikawa 1976.

Sakamoto Masako. 1990. "Diversification—The Case of Mitsui Bussan." In *General Trading Companies: A Comparative and Historical Study*, ed. Shin'ichi Yonekawa. Tokyo, United Nations University Press. Forthcoming.

Sasaki Junnosuke. 1979. *Dentōteki kōgyō gijutsu no taiyō* (*Modes of Traditional Mining Techniques*, HSDRJE-24/UNUP-99, 1981). UNU Research Paper No. HSDRJE-24J/UNUP-73. Tokyo, United Nations University Press.

———. 1980. *Nihon ni okeru zairai gijutsu to shakai* (*Endogenous Technology and Society in Japan*, HSDRJE-41/UNUP-221, 1981). UNU Research Paper No. HSDRJE-41J/UNUP-214. Tokyo, United Nations University Press.

———, ed. 1983. *Gijutsu no shakai-shi* (Social history of technology). 6 vols. Tokyo, Yuhikaku.

Sasama Hidefumi. 1981. *Seifun- seiyu-gyō no kindaika* (The modernization of the milling and seed-oil industry in Japan). UNU Research Paper No. HSDRJE-66J/UNUP-378. Tokyo, United Nations University Press.

Sato Mamoru. 1982. "Jitsugyō hoshū gakkō no seiritsu to tenkai" (Formation and development of vocational schools). In *Wagakuni ririku-ki no jitsugyō kyōiku. See* Toyoda 1982.

Sato Seizaburo. 1977. "Meiji Ishin no saikentō" (Re-examination of the Meiji Restoration). In *Kindai Nihon kenkyū nyūmon* (An introduction to the study of modern Japan). Tokyo, University of Tokyo Press.

Science and Technology Bureau, ed. 1980. *Kagaku-gijutsu no shi-teki tenkai* (Historical development of science and technology). Tokyo, Government Printing Office.

Seki Kiyohide, Taniuchi Tatsu, and Takahashi Man'emon. 1979. *Hokkaidō kaihatsu to gijutsu iten* (Development of Hokkaido and technology transfer). UNU Research Paper No. HSDRJE-22J/UNUP-71. Tokyo, United Nations University Press.

Seoka Makoto. 1982. *Zaibatsu keieisha to kirisutokyō shakai jigyōka* (Zaibatsu management and the Christian social worker). UNU Research Paper No. HSDRJE-74J/UNUP-398. Tokyo, United Nations University Press.

Shibuya Ryuichi. 1981. *Shiei shichiya no tenkai to seisaku taiō* (*The Emergence of Private Pawn Shops: Japanese Government Policy*, HSDRJE-59/UNUP-412, 1983). UNU Research Paper No. HSDRJE-59J/UNUP-371. Tokyo, United Nations University Press.

———. 1983. *Kōeki shichiya seidō no dōnyū to tenkai* (The public pawn-shop system: Introduction and development). UNU Research Paper No. HSDRJE-86J/UNUP-487. Tokyo, United Nations University Press.

Shinozuka Eiko. 1982. *Nihon no joshi rōdō* (Female labour in Japan). Tokyo, Toyo Keizai Shimpo Sha.

Shishido Toshio. 1970. *Mitsubishi shōji no kenkyū* (Study of the Mitsubishi Trading Company). Tokyo, Toyo Keizai Shimpo Sha.

Shoji Hikaru, and Miyamoto Ken'ichi. 1975. *Nippon no kōgai* (Pollution in Japan). Tokyo, Iwanami Shoten.

Shoji Yoshiro. 1982. *Ashio Dōzan kōgai jiken* (Mining pollution: The case of the Ashio Copper Mine). UNU Research Paper No. HSDRJE-77J/UNUP-401. Tokyo, United Nations University Press.

Shoji Yoshiro, and Sugai Masuro. 1984. *Tsūshi Ashio kōdoku jiken, 1877–1984* (General history of pollution incidents at the Ashio Copper Mine, 1877–1984). Tokyo, Shin'yo Sha.

Shukan Asahi, ed. 1981. *Nedan no Meiji, Taishō, Shōwa fūzoku shi* (Socio-cultural history of prices in the Meiji, Taisho, and Showa periods). Tokyo, Asahi Shimbun.

Song Yingxing. 1965. *Tien-Kung K'ai-Wu: Chinese Technology in the Seventeenth Century*. Trans. E-tu Zen Sun and Shiou-Chuan Sun. University Park, Penn., Pennsylvania State University Press.

Sugai Masuro. 1982. *Ashio Dōzan no kōgai mondai no tenkai katei* (*The Development Process of Mining Pollution at the Ashio Copper Mine*, HSDRJE-78/UNUP-416, 1983). UNU Research Paper No. HSDRJE-78J/UNUP-402. Tokyo, United Nations University Press.

Sumiya Mikio. 1952. *Nihon shokugyō kunren hattatsu shi* (A history of the development of Japanese vocational training). 2 vols. Tokyo, Nippon Rodo Kyokai.

Suzuki Tsuneo. 1990. "Development of Post-war General Trading Companies." In *General Trading Companies: A Comparative and Historical Study*, ed. Shin'ichi Yonekawa. Tokyo, United Nations University Press. Forthcoming.

Tada Hirokazu, ed. 1980. *A Selected Bibliography on Socio-economic Development of Japan. Part 1: Circa 1600–1940*. UNU Research Paper No. HSDRJE-26/UNUP-199. Tokyo, United Nations University Press.

Takada Hiroshi. 1978. *Kotoba no umi e* (Towards the sea of words). Tokyo, Shincho Sha.

Takai Toshio. 1980. *Watashi no jokō aishi* (My sad history of the female workers). Tokyo, Sodo Sha.

Takechi Kyozo. 1979. *Waga kuni botan sangyōshi no hitokoma* (Development of the button-making industry). UNU Research Paper No. HSDRJE-14J/UNUP-34. Tokyo, United Nations University Press.

———. 1980. *Waga kuni kakedokei seizō no tenkai to keitai* (The evolution and structure of the clock-making industry in Japan). UNU Research Paper No. HSDRJE-40J/UNUP-213. Tokyo, United Nations University Press.

Takeuchi Johzen. 1979. *Toshi-gata chūshō kōgyō no nōson kōgyōka* (*Agro-industrialization of Urban-Based Small Industries*, HSDRJE-15/UNUP-92, 1982). UNU Research Paper No. HSDRJE-15J/UNUP-35. Tokyo, United Nations University Press.

———. 1980. *Keiseiki no waga kuni jitensha sangyō* (*The Formation of the Japanese Bicycle Industry: A Preliminary Analysis of the Infrastructure of the Japanese Machine Industry*, HSDRJE-39/UNUP-241, 1981). UNU Research Paper No. HSDRJE-39J/UNUP-212. Tokyo, United Nations University Press.

Tamaki Akira. 1979. *Kangai shisutemu to chiiki nōgyō* (*Development of Local Culture and the Irrigation System of the Azusa Basin*, HSDRJE-4/UNUP-50, 1979). UNU Research Paper No. HSDRJE-4J/UNUP-24. Tokyo, United Nations University Press.

———. 1979. *Mizu no shisō* (Thoughts on water). Tokyo, Ronso Sha.

Tanaka Yoshio. 1980. *Kanazawa kinkō no keifu to hen'yō* (The transmission of metalsmithing down through the generations in Kanazawa and its transformation). UNU Research Paper No. HSDRJE-30J/UNUP-203. Tokyo, United Nations University Press.

Teranishi Juro. 1983. "Matsukata defurē no makuro keizai-teki bunseki" (Macroeconomic analysis of the Matsukata deflationary policy). In *Matsukata zaisei to shokusan kōgyō seisaku. See* Umemura and Nakamura 1983.

Tomosugi Takashi. 1979. *Keizai chikuseki no keitai to shakai henka* (Metamorphosis of economic accumulation and social change). UNU Research Paper No. HSDRJE-7J/UNUP-27. Tokyo, United Nations University Press.

———. 1980. *Tameike to shakai keisei* (Reservoirs and community formation: Reservoirs as a cultural factor). UNU Research Paper No. HSDRJE-45J/UNUP-218. Tokyo, United Nations University Press.

————. 1982. *Tochi no shōhinka to kahei no kigōka* (Modernization of the villagers' concept of money and land in the nineteenth century). UNU Research Paper No. HSDRJE-71J/UNUP-395. Tokyo, United Nations University Press.

————. 1983. *Shinkaichi ni okeru shakai keishiki to nōkyō* (Hokkaido's new frontier society and the development of agricultural co-operatives). UNU Research Paper No. HSDRJE-90J/UNUP-491. Tokyo, United Nations University Press.

Torii Yasuhiko. 1978. "Marayawata Purojekuto no keizai kōka" (Economic effects of the Malayawata Project). *Tekko-kai*, vol. 10.

Toyoda Toshio, ed. 1982. *Waga kuni ririku-ki no jitsugyō kyōiku* (Vocational education in Japan during the takeoff period). Tokyo, United Nations University Press.

————, ed. 1984. *Waga kuni sangyōka to jitsugyō kyōiku* (Japanese industrialization and vocational education). Tokyo, United Nations University Press.

Uchida Hoshimi. 1985. *Tokei-kōgyō no hatten* (Development of the Japanese watch industry). Tokyo, Nippon Keiei-shi Kenkyujo.

————. 1986. "Gijutsu seisaku no rekishi" (History of technology policy). In *Kindai Nihon no gijutsu to gijutsu seisaku. See* Uchida, Nakaoka, and Ishii 1986.

Uchida Hoshimi, Nakaoka Tetsuro, and Ishii Tadashi, eds. 1986. *Kindai Nihon no gijutsu to gijutsu seisaku* (Technology and technological policies in modern Japan). Tokyo, United Nations University Press.

Uchino Tatsuro. 1953. *Sengo Nippon keizai shi* (History of the Japanese post-war economy). Tokyo, Kodan Sha.

Ueda Tatsuzo. 1979. *Gankyō sangyō no hattatsu* (*The Development of the Eyeglass Industry in Japan*, HSDRJE-16/UNUP-98, 1979). UNU Research Paper No. HSDRJE-16J/UNUP-36. Tokyo, United Nations University Press.

————. 1980. *Jitensha sangyō no hattatsu* (*The Development of the Bicycle Industry in Japan after World War II*, HSDRJE-38/UNUP-224, 1981). UNU Research Paper No. HSDRJE-38J/UNUP-211. Tokyo, United Nations University Press.

Ui Jun. 1982. *Gijutsu dōnyū no shakai ni ataeta fu no shōgeki* (*The Negative Effects of Technology in Japan's Modernization Process: The Ashio Copper Mine Incident*, HSDRJE-84/UNUP-418, 1983). UNU Research Paper No. HSDRJE-84J/UNUP-408. Tokyo, United Nations University Press.

Umemura Mataji, and Nakamura Takafusa, eds. 1983. *Matsukata zaisei to shokusan kōgyō seisaku* (*Primary Industrial Policy in Japan: Crisis of a New Nation State*, forthcoming). Tokyo, United Nations University Press.

Wada Ei. 1983. *Tomioka nikki* (Tomioka diary). Tokyo, Chuo Koron Sha.

Watanabe Toshio. 1983. *Ajia suihei bungyō no jidai* (Horizontal division of labour in Asia). Tokyo, JETRO.

Yabu'uchi Kiyoshi. 1969. *Tenkō kaibutsu* (translation of *Tien-Kung K'ai-Wu*). Tokyo, Heibonsha.

Yagi Hironori. 1982. *Kurīku nōgyō no tenkai katei* (The development process of creek agriculture in the Saga Plains). UNU Research Paper No. HSDRJE-72J/UNUP-396. Tokyo, United Nations University Press.

Yamamoto Hirofumi. 1979. *Nihon no kōgyōka to yusō* (*Industrialization and Transportation in Japan*, HSDRJE-9/UNUP-90, 1981). UNU Research Paper No. HSDRJE-9J/UNUP-29. Tokyo, United Nations University Press.

————. 1980. *Tetsudō jidai no dōro yusō* (Road transportation in the age of railways). UNU Research Paper No. HSDRJE-34J/UNUP-207. Tokyo, United Nations University Press.

Yamamoto Shigemi. 1952. *Aa Nomugi Tōge* (Ah, Nomugi Pass [The sad history of the female silk-reeling workers]). 2 vols. Tokyo, Kadokawa Bunko.

Yamamoto Yuzo. 1983. "Okuma zaisei ron no hontai to gitai" (The theory and reality of Okuma's financial policy). In *Matsukata zaisei to shokusan kōgyō seisaku. See* Umemura and Nakamura 1983.

Yanagida Kunio. 1967. *Yanagida Kunio zenshū* (Collected works of Yanagida Kunio). 36 vols. Tokyo, Chikuma Shobo.

Yasaki Shunji. 1983. *Seisan soshiki no tenkai to suiden no shūraku-teki riyō* (Rice cultivation: Development of collective production systems). UNU Research Paper No. HSDRJE-89J/UNUP-490. Tokyo, United Nations University Press.

Yasuoka Shigeaki. 1981. *Dōzoku kigyō ni okeru shoyū to keiei* (*Ownership and Management of Family Businesses: An International Comparison*, HSDRJE-63/UNUP-414). UNU Research Paper No. HSDRJE-63J/UNUP-375. Tokyo, United Nations University Press.

Yokoyama Gennosuke. [1899] 1949. *Nihon no kasō shakai* (Lower strata of Japanese society). Tokyo, Iwanami Shoten.

Yonekawa Shin'ichi. 1981. *The Growth of Cotton-Spinning Firms and Vertical Integration: A Comparative Study of the UK, the USA, India, and Japan.* UNU Research Paper No. HSDRJE-49/UNUP-296. Tokyo, United Nations University Press.

———. 1983. *General Trading Companies: A Comparative and Historical Study.* UNU Research Paper No. HSDRJE-80/UNUP-404. Tokyo, United Nations University Press.

Yoshiki Fumio. 1979. *Kindai gijutsu dōnyū to kōzangyō no kindaika* (*How Japan's Metal Mining Industries Modernized.* HSDRJE-23/UNUP-83, 1980). UNU Research Paper No. HSDRJE-23J/UNUP-72. Tokyo, United Nations University Press.

Yuasa Mitsutomo. 1984. *Nihon no kagaku-gijutsu hyakunen-shi* (A 100-year history of science and technology in Japan). 2 vols. Tokyo, Chuo Koron Sha.

Collaborators

Technology and Urban Society
KOYANO SHOGO, Department of Human Relations, Tokiwa University, Ibaraki
ISHIZUKA HIROMICHI, Department of Social Sciences, Tokyo Metropolitan University, Tokyo
KURASAWA SUSUMU, Department of Social Sciences and Humanities, Tokyo Metropolitan University, Tokyo
KOSUGE NOBUHIKO, National Land Agency, Tokyo
TANAKA YOSHIO, Faculty of Economics, Kanazawa College of Economics, Ishikawa
NAKAMURA HACHIRO, Faculty of Education, Ibaraki University, Ibaraki
NAKAGAWA KIYOSHI, Department of Contemporary Society, Japan Women's University, Tokyo
HASHIMOTO TETSUYA, Department of Economics, Kanazawa University, Ishikawa
HAYASHI IKUO, Tokyo Metropolitan Government, Tokyo

Technology and Rural Society
TAMAKI AKIRA, Department of Economics, Senshu University, Tokyo
IMAMURA NARAOMI, Department of Agricultural Economics, University of Tokyo, Tokyo
HATATE ISAO, Department of Economics, Aichi University, Aichi
TOMOSUGI TAKASHI, Institute of Oriental Culture, University of Tokyo, Tokyo
NAGATA KEIJURO, Department of Agriculture, Shimane University, Shimane
SHICHINOHE CHOSEI, Department of Agricultural Economics, Hokkaido University, Hokkaido

Railways and Iron and Steel Industry
IIDA KEN'ICHI, Faculty of Engineering, Tokyo Engineering University, Tokyo
NAKURA BUNJI, Department of Social Science, Ibaraki University, Ibaraki
YAMAMOTO HIROFUMI, Department of Economics, Hosei University, Tokyo
AOKI EIICHI, Faculty of Education, Tokyo Gakugei Daigaku, Tokyo
ISHII ICHIRO, Department of Civil Engineering, Toyo University, Tokyo
HARADA KATSUMASA, Department of Economics, Wako University, Tokyo
MASUDA HIROMI, Chief librarian, Bunkyo University, Saitama

Textile Industries
YONEKAWA SHIN'ICHI, Department of Commerce, Hitotsubashi University, Tokyo
KIYOKAWA YUKIHIKO, Institute of Economic Research, Hitotsubashi University, Tokyo
KATO KOZABURO, Department of Economics, Senshu University, Tokyo
IZUMI TAKEO, Department of Economics, Senshu University, Tokyo

Small-scale Industries
KIKUURA SHIGEO, Department of Economics, Toyo University, Tokyo
UEDA TATSUZO, Department of Sociology, Kansai University, Osaka
TAKECHI JOHZEN, Department of Economics, Hiroshima University, Hiroshima
TAKECHI KYOZO, Faculty of Economics and Business Administration, Kinki University, Osaka

Mining Industries
SASAKI JUNNOSUKE, Faculty of Social Studies, Hitotsubashi University, Tokyo
YOSHIKI FUMIO, Sendai Daiichi High School, Miyagi
MURAKUSHI NISABURO, Department of Economics, Hosei University, Tokyo
KASUGA YUTAKA, Department of Economics and Business Administration, Chiba University, Chiba

Technology Transfer and Financial Institutions
SHIBUYA RYUICHI, Faculty of Economic Science, Komazawa University, Tokyo
ASAJIMA SHOICHI, Department of Business Administration, Senshu University, Tokyo
ASAI YOSHIO, Department of Economics, Seijo University, Tokyo
SAITO TOSHIHIKO, Department of Commerce, Chiba University of Commerce, Chiba
CHIBA OSAMU, Ministry of Agriculture, Tokyo
NAMIGATA SHOICHI, Department of Economics, Dokkyo University, Saitama
MUKAI YURIO, Tohoku University, Miyagi

Education and Vocational Training
TOYODA TOSHIO, Department of International Studies, Tokyo International University, Tokyo
SATO MAMORU, Faculty of Education, Akita University, Akita
IWAUCHI RYOICHI, School of Business Administration, Meiji University, Tokyo
TAKAGUCHI AKIHISA, Faculty of Education, University of Tottori, Tottori
TAKEUCHI JOHZEN, Department of Economics, Hiroshima University, Hiroshima
HANADA ARATA, Department of Sociology, Meiji Gakuin University, Tokyo
YAMAGISHI HARUO, Faculty of Education, Oita University, Oita
YAMASHITA EIICHI, Aichi Nishio High School, Aichi

Technology and Pollution
UI JUN, Department of Economics, University of Okinawa, Okinawa
IIJIMA NOBUKO, Department of Sociology, St. Andrew's University, Osaka
SHOJI YOSHIRO
SUGAI MASURO, Faculty of Economics, Kokugakuin University, Tokyo
HOSHINO YOSHIRO, Department of Economics, Teikyo University, Tokyo

Technology and Business Management
YASUOKA SHIGEAKI, Faculty of Commerce, Doshisha University, Kyoto
ISHIKAWA KENJIRO, Faculty of Commerce, Doshisha University, Kyoto
IWASHITA MASAHIRO, Faculty of Commerce, Doshisha University, Kyoto
KOBASHI ICHIRO, Department of Law, Doshisha University, Kyoto
SEOKA MAKATO, Faculty of Liberal Arts, Shiga University, Shiga
CHIMOTO AKIKO, Faculty of Commerce, Doshisha University, Kyoto
FUJITA TEIICHIRO, Faculty of Commerce, Doshisha University, Kyoto
NAKAMURA KEN, Institute for the Study of Humanities and Social Sciences, Doshisha
   University, Kyoto

Economic Policy
NAKAMURA TAKAFUSA, Department of Home Life Administration, Ochanomizu
   University, Tokyo
UMEMURA MATAJI, Department of Economics, Soka University, Tokyo
ABE TAKESHI, Department of Economics, Osaka University, Osaka
SAITO OSAMU, Institute of Economic Research, Hitotsubashi University, Tokyo
TERANISHI JURO, Institute of Economic Research, Hitotsubashi University, Tokyo
BANNO JUNJI, Institute of Social Science, University of Tokyo, Tokyo
INOKI TAKENORI, Department of Economics, Osaka University, Osaka
YAMAMOTO YUZO, Research Institute for Humanistic Studies, Kyoto University,
   Kyoto
MIKURIYA TAKASHI, Department of Politics, Tokyo Metropolitan University, Tokyo
MUROYAMA MASAYOSHI, Department of Political Science, Takushoku University,
   Tokyo

Economic Thought
CHO YUKIO, President, Tokyo University of Foreign Studies, Tokyo
SHODA KEN'ICHIRO, Department of Political Science, Waseda University, Tokyo
NAKAMURA HIDEICHIRO, Faculty of Business Administration, Tama University,
   Tokyo
FUKUSHIMA SHINGO, Department of Law, Senshu University, Tokyo
TAMAKI AKIRA, Senshu University, Tokyo

Female Labour and Technology
NAKAMURA MASANORI, Faculty of Economics, Hitotsubashi University, Tokyo
KASE KAZUTOSHI, Department of Resource Management, Tokyo University of
   Fisheries, Tokyo
CORRADO MOLTEIN, Faculty of Social Studies, Hitotsubashi University, Tokyo
SHIOTA SAKIKO, Department of Economics, Takasaki City University of Economics,
   Gumma
NISHINARITA YUTAKA, Faculty of Economics, Hitotsubashi University, Tokyo
MIYAKE AKIMASA, Faculty of Liberal Arts, Chiba University, Chiba

General Trading Companies and Technology Transfer
YONEKAWA SHIN'ICHI, Department of Commerce, Hitotsubashi University, Tokyo
KAWABE NOBUO, Faculty of Integrated Arts and Sciences, Hiroshima Univeristy,
   Hiroshima
SAKAMOTO MASAKO, Department of Consumer Economics, Nagoya Economics Uni-
   versity, Aichi

Suzuki Tsuneo, Department of Economics, Kurume University, Fukuoka
Maeda Kazutoshi, Faculty of Business Management, Komazawa University, Tokyo

History of Technology
Uchida Hoshimi, Department of Business Administration, Tokyo Keizai Daigaku, Tokyo
Ishii Tadashi, The Patent Agency, Tokyo
Nakaoka Tetsuro, Faculty of Economics, Osaka City University, Osaka
Sato Mitsuru, Faculty of Economics, Osaka City University, Osaka
Shiozawa Yoshinori, Faculty of Economics, Osaka City University, Osaka
Honda Kenkichi, Faculty of Economics, Osaka City University, Osaka

Technology in Changing Life-styles
Otsuka Tsutomu, Toita Women's College, Tokyo
Sasama Hidefumi, Institute of Industrial History, Tokyo
Nakagome Shozo, Bunka Fukuso University, Tokyo

Technology Transfer and the Development of Hokkaido
Seki Kiyohide, Faculty of Letters, Hokkaido University, Hokkaido
Kikuchi Hiroaki, Department of Architectural Engineering, Hokkaido Institute of Technology, Hokkaido
Takahashi Man'emon, Faculty of Agriculture, Hokkaido University, Hokkaido
Taniuchi Toru, Faculty of Education, University of Tokyo, Tokyo
Miki Tsuyoshi, Department of Medicine, Sapporo Medical College, Hokkaido
Yamagishi Takashi, Hokkaido Research Laboratory of Public Health, Hokkaido
Murayama Etsuo, Faculty of Engineering, Hokkaido University, Hokkaido

Area Study: Niigata
Ikeda Shoji, Faculty of Education, Niigata University, Niigata
Kagoshima Gihei, Niigata-Tsubame High School, Niigata
Kato Shinzo, Japan Metal Tableware Exporters' Association, Niigata
Saito Yoshinobu, Niizu High School, Niigata
Suzuki Hayao, Numata High School, Niigata
Suzuki Masakiyo, Tsubame High School, Niigata
Sotoyama Noboru, Sanjo Metalists' Youth Club, Niigata
Wakatsuki Takeo, Sanjo City Library, Niigata

Other
Kawakami Takeshi, Kikusaka Clinic, Tokyo
Kambayashi Shigenobu, Yanagihara Hospital, Tokyo
Namie Ken, Japan Library Association, Tokyo

Index

276